# ECOLOGICAL RESILIENCE
Response to Climate Change
and Natural Disasters

# ECOLOGICAL RESILIENCE
## Response to Climate Change and Natural Disasters

*Edited by*
**Kimberly Etingoff**

| Apple Academic Press Inc. | Apple Academic Press Inc. |
| 3333 Mistwell Crescent | 9 Spinnaker Way |
| Oakville, ON L6L 0A2 | Waretown, NJ 08758 |
| Canada | USA |

©2016 by Apple Academic Press, Inc.

First issued in paperback 2021

*Exclusive worldwide distribution by CRC Press, a member of Taylor & Francis Group*

No claim to original U.S. Government works

ISBN 13: 978-1-77463-600-8 (pbk)
ISBN 13: 978-1-77188-310-8 (hbk)

---

**Library and Archives Canada Cataloguing in Publication**

---

Ecological resilience : response to climate change and natural disasters / edited by Kimberly Etingoff.

Includes bibliographical references and index.
Issued in print and electronic formats.
ISBN 978-1-77188-310-8 (hardcover).--ISBN 978-1-77188-311-5 (pdf)
1. Resilience (Ecology). 2. Human ecology. 3. Climatic changes--Social aspects.
4. Natural disasters--Social aspects. 5. City planning--Environmental aspects--Case studies.
6. Public health--Environmental aspects--Case studies.  I. Etingoff, Kim, editor

| GF21.E36 2016 | 304.2 | C2016-900321-3 | C2016-900322-1 |

---

**Library of Congress Cataloging-in-Publication Data**

---

Names: Etingoff, Kim.
Title: Ecological resilience : response to climate change and natural disasters / [edited by] Kimberly Etingoff.
Description: Toronto : Apple Academic Press, 2016. | Includes bibliographical references and index.
Identifiers: LCCN 2016001419 (print) | LCCN 2016007320 (ebook) | ISBN 9781771883108 (hardcover : alk. paper) | ISBN 9781771883115 ()
Subjects: LCSH: Resilience (Ecology) | Emergency management. | Climate change mitigation.
Classification: LCC QH75 .E2725 2016 (print) | LCC QH75 (ebook) | DDC 333.95/16--dc23
LC record available at http://lccn.loc.gov/2016001419

---

Apple Academic Press also publishes its books in a variety of electronic formats. Some content that appears in print may not be available in electronic format. For information about Apple Academic Press products, visit our website at **www.appleacademicpress.com** and the CRC Press website at **www.crcpress.com**

# About the Editor

---

**KIMBERLY ETINGOFF**

Kim Etingoff has a master's degree in urban and environmental policy and planning from Tufts University. Her recent experience includes researching a report on food resiliency within the city of Boston with Initiative for a Competitive Inner City. She worked in partnership with Dudley Street Neighborhood Initiative and Alternatives for Community and Environment to support a community food-planning process based in a Boston neighborhood, which was oriented toward creating a vehicle for community action around urban food issues, providing extensive background research to ground the resident-led planning process. She has worked in the Boston Mayor's Office of New Urban Mechanics, and has also coordinated and developed programs in urban agriculture and nutrition education. In addition, she has many years of experience researching, writing, and editing educational and academic books on environmental and food issues.

# Contents

# Acknowledgment and How to Cite

The editor and publisher thank each of the authors who contributed to this book. The chapters in this book were previously published elsewhere. To cite the work contained in this book and to view the individual permissions, please refer to the citation at the beginning of each chapter. Each chapter was carefully selected by the editor; the result is a book that looks at ecological resilience from a variety of perspectives. The chapters included are broken into four sections, which describe the following topics:

- The authors of Chapter 1 introduce an evaluative model connecting various policy scenarios to resilience outcomes in urban areas, in terms of human activity and environmental effects.
- Chapter 2 seeks to understand the complexity of creating sustainable and resilient cities by incorporating vulnerability and transition theory into resilience theory.
- In order to translate abstract concepts of resilience into practical applications, the authors of Chapter 3 identify multiple attributes that promote system resilience and help practitioners assess the current and potential resilience of relevant systems.
- Chapter 4 understands community action as a key component of achieving sustainable and resilient development, and identifies the ways in which community vitality supports resilience using research conducted in thirty-five Canadian communities.
- The author of Chapter 5 uses the case study of the Danungdafu Forestation Area in Taiwan to explore the capacities and challenges of managing resilience across multiple scales and levels within governance systems, and concludes that more participatory and just governance is needed to achieve greater resilience.
- Chapter 6 introduces the concept of a Dismantable City as a way to make urban areas more adaptable and resilient in the face of climate change effects. It specifically lays out ways that cities can adjust in terms of multilayer urbanism, light urbanism, and transformable urbanism.
- The authors of Chapter 7 suggest that because of the complexity of urban areas and the uncertainties involved in hazard evaluation, practitioners and researchers should first conduct uncertainty analyses of urban systems.

- Chapter 8 connects energy infrastructure to natural disasters in Japan, and proposes six criteria that allow energy systems to enhance societal resilience after natural disasters.
- After defining what health resilient communities look like, Chapter 9 discusses three examples of existing public health programs and their capacity to improve community health resiliency.
- The authors of Chapter 10 describe a projection model and risk management framework used by the U.S. Centers for Disease Control and Prevention to improve public health adaptation within communities.
- Chapter 11 presents a Los Angeles-based public health training program meant to improve community disaster resilience, and lays the groundwork for an evaluative research project involving this program.

# List of Contributors

**B. Barroca**
Lab'Urba, Paris-Est university, 5 Bd Descartes, 77454 Marne-La-Vallée, France

**P. Bernardara**
EDF Energy R&D UK Centre, London, UK

**Anita Chandra**
RAND Corporation, 1200 South Hayes Street, Arlington, VA 22202, USA

**Ralph Chapman**
School of Geography, Environment, and Earth Sciences, Victoria University of Wellington, PO BOX 600, Wellington 6242, New Zealand

**Ann Dale**
School of Environment and Sustainability, Royal Roads University, 2005 Sooke Road, Victoria, BC, V9B 5Y2, Canada

**Kristie L. Ebi**
Department of Medicine, Stanford University, 260 Panama Street, Stanford, CA 94305, USA

**David Eisenman**
Los Angeles County Department of Public Health Emergency Preparedness and Response Program, 600 S. Commonwealth Avenue, Suite 700, Los Angeles, CA 90005, USA and Center for Public Health and Disasters, UCLA Fielding School of Public Health, P.O. Box 951772, Los Angeles, CA 90095, USA

**Hooman Farzaneh**
Graduate School of Energy Science, Kyoto University, Yoshida honmachi, Sakyo-ku, Kyoto 606-8501, Japan

**Stella Fogleman**
Los Angeles County Department of Public Health Emergency Preparedness and Response Program, 600 S. Commonwealth Avenue, Suite 700, Los Angeles, CA 90005, USA

**S. Girard**
Team Mistis, Inria Grenoble Rhône-Alpes & LJK, Inovallée, 655, av. de l'Europe, Montbonnot, 38334 Saint-Ismier CEDEX, France

**Astrid Hendricks**
Los Angeles County Department of Public Health Emergency Preparedness and Response Program, 600 S. Commonwealth Avenue, Suite 700, Los Angeles, CA 90005, USA

**Jeremy J. Hess**
Climate and Health Program, Division of Environmental Hazards and Health Effects, National Center for Environmental Health, Centers for Disease Control and Prevention, Atlanta, GA 30341, USA, Department of Environmental Health, Rollins School of Public Health at Emory University, Atlanta, GA 30322, USA, and Department of Emergency Medicine, School of Medicine, Emory University, Atlanta, GA 30322, USA

**Philippa Howden-Chapman**
New Zealand Centre for Sustainable Cities, University of Otago, 23A Mein Street, Newtown, Wellington 6242, New Zealand

**Keiichi N. Ishihara**
Graduate School of Energy Science, Kyoto University, Yoshida honmachi, Sakyo-ku, Kyoto 606-8501, Japan

**David A. Kerner**
The Tauri Group, LLC, 6361 Walker Lane, Suite 100, Alexandria, VA 22310, USA

**Chris Ling**
School of Environment and Sustainability, Royal Roads University, 2005 Sooke Road, Victoria, BC, V9B 5Y2, Canada

**George Luber**
Climate and Health Program, Division of Environmental Hazards and Health Effects, National Center for Environmental Health, Centers for Disease Control and Prevention, Atlanta, GA 30341, USA

**Aizita Magana**
Los Angeles County Department of Public Health Emergency Preparedness and Response Program, 600 S. Commonwealth Avenue, Suite 700, Los Angeles, CA 90005, USA

**Gino D. Marinucci**
Climate and Health Program, Division of Environmental Hazards and Health Effects, National Center for Environmental Health, Centers for Disease Control and Prevention, Atlanta, GA 30341, USA

**G. Mazo**
Team Mistis, Inria Grenoble Rhône-Alpes & LJK, Inovallée, 655, av. de l'Europe, Montbonnot, 38334 Saint-Ismier CEDEX, France

**Benjamin McLellan**
Graduate School of Energy Science, Kyoto University, Yoshida honmachi, Sakyo-ku, Kyoto 606-8501, Japan

**Lenore Newman**
School of Environment and Sustainability, Royal Roads University, 2005 Sooke Road, Victoria, BC, V9B 5Y2, Canada

**Alonzo Plough**
Robert Wood Johnson Foundation, Route 1 and College Road East, P.O. Box 2316, Princeton, NJ 08543, USA

**Edward Randal**
New Zealand Centre for Sustainable Cities, University of Otago, 23A Mein Street, Newtown, Wellington 6242, New Zealand

**Rob Roggema**
Van Hall Larenstein University of Applied Sciences, Velp 6880 GB, The Netherlands and National Institute for Design Research, Swinburne University of Technology, Hawthorn, VIC 3122, Australia

**Shubhayu Saha**
Climate and Health Program, Division of Environmental Hazards and Health Effects, National Center for Environmental Health, Centers for Disease Control and Prevention, Atlanta, GA 30341, USA

**Leanne Seeliger**
Human Sciences Research Council, Economic Performance and Development; Private Bag X9182, Cape Town 8000, South Africa

**Hsing-Sheng Tai**
Department of Natural Resources and Environmental Studies, National Dong Hwa University, No.1, Sec.2, Da Hsueh Rd., Shoufeng, Hualien 97401, Taiwan

**Jennifer Tang**
Center for Health Services and Society, David Geffen School of Medicine, 10920 Wilshire Boulevard, Suite 300, Los Angeles, CA 90024, USA

**J. Scott Thomas**
Stetson Engineers Inc., 2171 E. Francisco Blvd., Suite K, San Rafael, CA 94901, USA and University College, University of Denver, 2211 South Josephine St., Denver, CO 80208, USA

**Ivan Turok**
Human Sciences Research Council, Economic Performance and Development; Private Bag X9182, Cape Town 8000, South Africa

**Christopher K. Uejio**
Climate and Health Program, Division of Environmental Hazards and Health Effects, National Center for Environmental Health, Centers for Disease Control and Prevention, Atlanta, GA 30341, USA and Department of Geography, Florida State University, 113 Collegiate Loop, Tallahassee, FL 32306, USA

**N. Agya Utama**
Graduate School of Energy Science, Kyoto University, Yoshida honmachi, Sakyo-ku, Kyoto 606-8501, Japan

**Ken Wells**
Center for Health Services and Society, David Geffen School of Medicine, 10920 Wilshire Boulevard, Suite 300, Los Angeles, CA 90024, USA

**Malcolm Williams**
RAND Corporation, 1776 Main Street, Santa Monica, CA 90401, USA

**Qi Zhang**
Graduate School of Energy Science, Kyoto University, Yoshida honmachi, Sakyo-ku, Kyoto 606-8501, Japan

**Pengjun Zhao**
New Zealand Centre for Sustainable Cities, University of Otago, 23A Mein Street, Newtown, Wellington 6242, New Zealand

# Introduction

Resilience is a growing and vital concept in a world increasingly faced with the devastating effects of climate change. Cities, rural areas, and entire regions must learn to adapt to natural disasters and long-term socioecological changes caused by a changing climate. This book presents some of the latest research on resilience strategies around the world. Research such as this is necessary to create new ideas and to evaluate established ones, in an effort to make communities more adaptable and to increase people's survival and quality of life while living with the reality of climate change.

Part I offers researchers' definitions of resilience, as well as various ways of measuring it, since resilience is still a concept in transition. Part II describes some general strategies for increasing communities' resilience at multiple levels. Parts III and IV dive into the specific dimensions of resilience, tying it to both energy infrastructure and systems and public health.

*Kimberly Etingoff*

The resilience of cities in response to natural disasters and long-term climate change has emerged as a focus of academic and policy attention. In particular, how to understand the interconnectedness of urban and natural systems is a key issue. Chapter 1, by Zhao and colleagues, introduces an urban model that can be used to evaluate city resilience outcomes under different policy scenarios. The model is the Wellington Integrated Land Use-Transport-Environment Model (WILUTE). It considers the city (i.e., Wellington) as a complex system characterized by interactions between a variety of internal urban processes (social, economic and physical) and the natural environment. It is focused on exploring the dynamic relations between human activities (the geographic distribution of housing and employment, infrastructure layout, traffic flows and energy consumption), environmental effects (carbon emissions, influences on local natural and ecological systems) and potential natural disasters (e.g., inundation due to sea level rise and storm events) faced under different policy scenarios. The model gives insights that are potentially useful for policy to enhance the city's resilience, by modelling outcomes, such as the potential for reduction in transportation energy use, and changes in the vulnerability of the city's housing stock and transport system to sea level rise.

Cities at all stages of development need to provide jobs, food and services for their people. There is no formula that can unilaterally be applied in all urban environments to achieve this. The complex interaction of social, economic and ecological cycles within cities makes it impossible to predict outcomes. Resilience theory, with its engineering, multi-equilibria and socio-ecological approaches, provides some of the foundations for understanding the full range of the complex social and ecological interactions that underpin sustainable cities. In Chapter 2, Seeliger and Turok propose that these insights could be extended by a sharper focus on the social and technological innovation that has traditionally been the emphasis of vulnerability and transition theories respectively.

If resilience theory is to be of practical value for policy makers and resource managers, the theory must be translated into sensible decision-support tools. In Chapter 3, Kerner and Thomas present a set of resilience attributes, developed to characterize human-managed systems, that helps system stakeholders to make practical use of resilience concepts in tangible applications. In order to build and maintain resilience, these stakeholders must be able to understand what qualities or attributes enhance—or detract from—a system's resilience. The authors describe standardized resilience terms that can be incorporated into resource management plans and decision-support tools to derive metrics that help managers assess the current resilience status of their systems, make rational resource allocation decisions, and track progress toward meeting goals. Their intention is to provide an approachable set of terms for both specialists and non-specialists alike to apply to programs that would benefit from a resilience perspective. These resilience terms can facilitate the modeling of resilience behavior within systems, as well as support those lacking access to sophisticated models. The goal is to enable policy makers and resource managers to put resilience theory to work in the real world.

Community level action towards sustainable development has emerged as a key scale of intervention in the effort to address our many serious environmental issues. This is hindered by the large-scale destruction of both urban neighbourhoods and rural villages in the second half of the twentieth century. Communities, whether they are small or large, hubs of experimentation or loci of traditional techniques and methods, can be said to have a level of community vitality that acts as a site of resilience, adaptation and innovation in the face of environmental challenges. Chapter 4, by Dale and colleagues, outlines how community vitality acts as a cornerstone of sustainable development and suggests some courses for future research. A meta-case analysis of thirty-five Canadian communities reveals the characteristics of community vitality emerging from sustainable development experiments and its relationship to resilience, applied specifically to community development.

Resilience thinking has strongly influenced how people understand and pursue sustainability of linked social-ecological systems. Resilience thinking highlights the need to build capacity and manage general system properties in a complex, constantly changing world. In chapter 5, Tai modified an analytical framework to address associations among cross-scale and cross-level dynamics, attributes of governance, and capacity to enhance resilience. The Danungdafu Forestation Area represents one of Taiwan's most controvisal cases concerning land use, indigenous rights, and environmental issues. Analysis of this Taiwanese experience from a social-ecological perspective can show how current capacities for managing resilience are related to critical governance attributes. Analysis helped identify fundamental flaws in current governance and key issues needing to be addressed. The Danungdafu Forestation Area should transition towards a governance regime that is more participatory, deliberative, multi-layered, accountable, just, and networked. This can be done by developing an intermediate level institution that coordinates the cross-scale and cross-level interactions that better fit this social-ecological system.

When we use the urban metabolism model for urban development, the input in the model is often valuable landscape, being the resource of the development, and output in the form of urban sprawl, as a result of city transformations. The resilience of these "output" areas is low. The lack of resilience is mainly caused by the inflexibility in these areas where existing buildings, infrastructure, and public space cannot be moved when deemed necessary. In Chapter 6, Roggema proposes a new vision for the city in which the locations of these objects are flexible and, as a result, the resilience is higher: a Dismantable City. Currently, the development of this sort of city is constrained by technical, social, and regulatory practice. However, the perspective of a Dismantable City is worthwhile because it is able to deal with sudden, surprising, and unprecedented climate impacts. Through self-organizing processes the city becomes adjustable and its objects mobile. This allows the city to configure itself according to environmental demands. The city is then able to withstand or even anticipate floods, heat waves, droughts, or bushfires. Adjustability can be found in several directions: creating multiple layers for urban activities (multi-layer urbanism), easing the way objects are constructed (light urbanism), or reusing abandoned spaces (transformable urbanism).

Chapter 7, by Barroca and colleagues, describes how urbanization has led to a higher concentration of both persons and property, which increases the potential degree of damage liable to occur in crisis situations. Urban areas have become increasingly complex socio-technical systems where the inextricable tangle of activities, networks and regions means disruptions propagate rather than disseminate. In risk anticipation, measures of prevention and anticipation are generally defined by using hazard modelling. The relevance of this approach may be subject to discussion (Zevenbergen et al., 2011) particularly in view of the large number of uncertainties that make hazard evaluation so difficult. For this reason, uncertainty analysis is initially called upon in a theoretical approach before any applied approach. Generally, the uncertainty under study is not assessed in hydrological studies. This uncertainty is related to the choice of evaluation model used for extreme values. This application has been used on the territory of the town of Besançon in eastern France. Strategic orientations for regional resilience are presented taking into account the high levels of uncertainty concerning estimates for possible flow rates.

The natural and subsequent human disasters of March 11, 2011 in Japan have brought into focus more than ever the importance of resilience and risk mitigation in the construction of energy infrastructure. Chapter 8, by McLellan and colleagues, introduces some of the critical issues and discusses the implications of energy in alleviating or exacerbating the risks of natural disasters. Additionally, it presents a framework for considering the risks of energy systems from a broad perspective. The connection is drawn between design for sustainability and the risks associated with energy systems in natural disasters. As a result of the assessment, six criteria are proposed for energy systems to contribute to societal resilience in the face of natural disasters—they should be: (1) Continuous; (2) Robust; (3) Independent; (4) Controllable; (5) Non-hazardous; and (6) Matched to demand.

Current public health strategies, policies, and measures are being modified to enhance current health protection to climate-sensitive health outcomes. These modifications are critical to decrease vulnerability to climate variability, but do not necessarily increase resilience to future (and different) weather patterns. Communities resilient to the health risks of climate change anticipate risks; reduce vulnerability to those risks; prepare for and respond quickly and effectively to threats; and recover faster, with increased capacity to prepare for and respond to the next threat. Increasing resilience includes top-down (e.g., strengthening and maintaining disaster risk management programs) and bottom-up (e.g., increasing social capital) measures, and focuses not only on the risks presented by climate change but also on the underlying socioeconomic, geographic, and other vulnerabilities that affect the extent and magnitude of impacts. In Chapter 9, Ebi and colleagues discuss three examples of public health programs designed for other purposes that provide opportunities for increasing the capacity of communities to avoid, prepare for, and effectively respond to the health risks of extreme weather and climate events. Incorporating elements of adaptive management into public health practice, including a strong and explicit focus on iteratively managing risks, will increase effective management of climate change risks.

Climate change is anticipated to have several adverse health impacts. Managing these risks to public health requires an iterative approach. As with many risk management strategies related to climate change, using modeling to project impacts, engaging a wide range of stakeholders, and regularly updating models and risk management plans with new information—hallmarks of adaptive management—are considered central tenets of effective public health adaptation. The Centers for Disease Control and Prevention has developed a framework, entitled Building Resilience Against Climate Effects, or BRACE, to facilitate this process for public health agencies. Its five steps are laid out in Chapter 10, by Marinucci and colleagues. Following the steps laid out in BRACE will enable an agency to use the best available science to project likely climate change health impacts in a given jurisdiction and prioritize interventions. Adopting BRACE will also reinforce public health's established commitment to evidence-based practice and institutional learning, both of which will be central to successfully engaging the significant new challenges that climate change presents.

Public health officials need evidence-based methods for improving community disaster resilience and strategies for measuring results. Chapter 11, by Eisenman and colleagues, describes how one public health department is addressing this problem. This paper provides a detailed description of the theoretical rationale, intervention design and novel evaluation of the Los Angeles County Community Disaster Resilience Project (LACCDR), a public health program for increasing community disaster resilience. The LACCDR Project utilizes a pretest–posttest method with control group design. Sixteen communities in Los Angeles County were selected and randomly assigned to the experimental community resilience group or the comparison group. Community coalitions in the experimental group receive training from a public health nurse trained in community resilience in a toolkit developed for the project. The toolkit is grounded in theory and uses multiple components to address education, community engagement, community and individual self-sufficiency, and partnerships among community organizations and governmental agencies. The comparison communities receive training in traditional disaster preparedness topics of disaster supplies and emergency communication plans. Outcome indicators include longitudinal changes in inter-organizational linkages among community organizations, community member responses in table-top exercises, and changes in household level community resilience behaviors and attitudes. The LACCDR Project is a significant opportunity and effort to operationalize and meaningfully measure factors and strategies to increase community resilience. This paper is intended to provide public health and academic researchers with new tools to conduct their community resilience programs and evaluation research. Results are not yet available and will be presented in future reports.

# PART I

# DEFINING AND MEASURING RESILIENCE

## CHAPTER 1

# Understanding Resilient Urban Futures: A Systemic Modelling Approach

PENGJUN ZHAO, RALPH CHAPMAN, EDWARD RANDAL, AND PHILIPPA HOWDEN-CHAPMAN

## 1.1 INTRODUCTION

Today, more than 95% of the world's population lives in less than 10% of the Earth's land area, mainly in cities and towns. The level of urbanization continues to rise, and it is forecast that by 2050, the urban population could be 6.29 billion, which will account for 69% of the total global population [1]. The most important influences of urbanization on the environment are energy use and the related increase in the emission of greenhouse gases, due to changes in land use and urban human activities [2,3]. The human population of the planet has increased four-fold over the last one hundred years, while—in the same time period—material and energy use has increased ten-fold [1]. With increasing urbanization, cities now consume about 75% of total global energy and produce 80% of its greenhouse gases [1].

Cities have become significant players in regard to policies, which are attempting to respond to peak oil and climate change. Recently, these policies have been focused on building resilient cities, which aim to enhance a city's ability to respond to a natural resource shortage and the recognition of the human impact on climate change [4]. Resilient cities are believed to adapt better to change through adjusting inner systems, for example, by changing their transport-land use system to reduce energy consumption and exposure of the system to potential natural disasters (e.g., sea-level rise). A resilient city reduces its ecological footprint (e.g., energy consumption), while simultaneously improving its quality of life. Resilient city policies are concerned with strengthening a city's capacity to adapt to shocks, such as natural disasters [5]. Such policies increase the degree of collaboration between urban subsystems (social, environmental-infrastructural, economic and institutional systems), while enhancing the robustness of each subsystem.

In particular, it is vital to build resilient urban futures for coastal cities. Coastal cities play a crucial role in human social and economic development in the world. Most global cities, such as London, New York, Sydney, Amsterdam, Tokyo, Hong Kong and Shanghai, are coastal cities. Thirteen of the world's 20 megacities are situated along coastlines, and more than two-thirds of the world's large cities are in coastal areas vulnerable to global warming and rising sea levels. In the 20th century, sea levels rose by an estimated average of 17 cm, and global mean projections for sea level rise between 1990 and 2080 range from 22 cm to 34 cm, according to reports by the IPCC (the Intergovernmental Panel on Climate Change) [6]. However, a recent research finding showed that sea-level rise was 3.2 ± 0.5 mm per year during the period from 1993 to 2011, which is 60% faster than the best IPCC estimate of 2.0 mm per year for the same period [7]. The sea level rise could range from 37 to 60 cm between 2000 and 2100, according to a high model scenario. The low elevation coastal zone—the continuous area along coastlines that is less than 10 m above sea level—represents 2% of the world's land area, but contains 10% of its total population and 13% of its urban population [8]. There are 3351 cities in the low elevation coastal zones around the world [8]. Urban areas are most vulnerable to sea level rise, and few coastal cities are likely to be spared by climate change.

A transport system is a precondition for social and economic activities in a city, enabling passenger and goods movements. Transport systems have been attracting the attention of the public, politicians and planners in building resilient cities [9]. One reason for this is that transportation is the fastest growing contributor to global climate change and urban health problems in the past few decades. Global transport emissions contributed an estimated 22% of direct $CO_2$ emissions in 2010, and 75% of global transport emissions were due to road transport [10]. The share is expected to continue growing at a rate of 1.7% per year up to 2030 [11]. In particular, total emissions have increased continuously for passenger transport (an increase of 27% between 1990 and 2004) [12]. Total vehicle miles travelled (VMT) is still growing globally, even though the growth seems to be slowing in several developed countries [13,14]. Motor vehicles can also cause the emissions of other environmentally harmful gases, such as NOx, $SO_2$ and particulate matter [15]; so, abating carbon yields substantial co-benefits. Urban air pollution caused by transport and traffic injuries combined together kill about 2.5 million people every year [16,17]. The other reason why transport systems are important to sustainability and resilience is that transport systems are often criticized for having much less adaptive capacity than other city systems. Once transport infrastructures are built, in particular, airports, ports, railways, highways and main roads, they are hard to change. Transport emissions are affected by many factors in city systems, for example, land-use patterns, planning constraints, city and transport network design, public transit services, parking policies, vehicle and fuel technologies and other factors related to individual travel behaviours [18]. Accordingly, many policies have been used in an attempt to change these systems and reduce transport emissions. However, these policies are still often criticized for their inefficiencies. Apart from the limitations of individual policies, the lack of integration and misalignment between these individual policies is a major reason for the criticisms. Cities are complex systems [19,20]. A systemic solution, which takes all these factors into account, would, in principle, be a more efficient way of reducing emissions from transport [18].

In addition, many policies designed to reduce GHG (Greenhouse Gas) emissions from road transport are focused on vehicle and fuel technology. However, many transportation researchers argue that individual travel behaviour is a critical aspect of sustainable transportation and just as important as technical factors and infrastructure supply. Many empirical studies have already provided evidence for this [21]. Therefore, developing a model based on individuals' travel behaviour is necessary in order to evaluate GHG emissions reduction policies.

This paper introduces an urban model that can be used to evaluate city resilience outcomes under a range of policy scenarios. The model is the Wellington Integrated Land Use-Transport-Environment Model (WILUTE), which is currently being developed by the New Zealand Centre for Sustainable City, University of Otago. The model is used to consider different policy scenarios and assess resilience. In the model, resilience is measured in three aspects. One is a city's capacity to reduce energy consumption and GHG emissions, in particular, from transportation changes. Another is the vulnerability of a city's land use and transport system to sea-level rise. The other is the costs related to reduce the vulnerability to a safe level with a consideration of its financial capacity. If the vulnerability and the costs are too high for its financial capacity, the city has a low resilience. Wellington is a typical small to medium-sized (city region population around 490,000) coastal city. It is the capital of New Zealand, which, with most cities being coastal, is vulnerable to sea-level rise, but resilient in terms of institutional, policy and human capacity. A model based on Wellington is useful in illustrating climate mitigation and resilience policies for other medium-sized coastal cities in New Zealand and in other countries.

The WILUTE model considers the city as a complex system characterized by interactions between a variety of internal urban processes (social, economic and physical) and the natural environment. It is focused on exploring the dynamic relations between human activities (the geographic distribution of housing and employment, infrastructure layout, traffic flows and energy consumption), environmental effects (carbon emissions, influences on local natural and ecological systems) and potential natural disasters (e.g., inundation due to sea level rise and storm events) faced under different policy scenarios. The model gives insights that are potentially useful for policy to enhance the city's resilience, by modelling key outcomes, such as traffic flows, transport energy consumption, GHG emissions, distribution of houses and commercial areas and traffic links. These key outcomes of modelling are the main factors influencing the changes in a city's resilience, as indicated by the aspects described above (transportation emissions response, etc.)

## 1.2 THE CITY AS A SYSTEM

The city is a complex system characterized by nonlinear behaviour, self-organization and emergent properties [19,20]. It is permeated by uncertainty and discontinuities [22]. The city as a whole is far from equilibrium and is more than the sum of its subsystems. Urban development is a complex process, involving a wide range of activities, actors and policies on a variety of geographical and administrative scales (country, state, regional, municipal and community). The urban change process consists of many dynamic sub-processes, such as economic, social, spatial, cultural and institutional processes. It involves a variety of city activities and sectors. For example, the urban spatial change process involves urban development and redevelopment activities, urban planning and design, household residential location choice, urban governance, transport demand and supply, industrial and commercial firms' location choice, changes in the technical sophistication of building and transport technology, etc. There is a high level of interaction between these sub-processes (Figure 1).

The urban change process has effects on urban sustainability through impacts on human well-being and ecosystems. As mentioned above, the major output of an urban system is the discharge of waste and emissions into the biosphere. Urban areas, in this sense, are primarily sites of consumption of water, energy, food, materials, land and other natural resources, and the discharges that reflect this consumption have many health effects. For example, urban outdoor air pollution contributes to approximately 5% of trachea, bronchus and lung cancer, 2% of cardiorespiratory mortality and about 1% of respiratory infections in the world in 2001 [23].

While the urban change process is too complex and organic to be fully optimized, the process can be reorganized and improved, for example, to reduce the discharge of waste and emissions into the biosphere and the consequent impacts on well-being. There are two strategies that are usually used to reduce the negative effects of urban development on emissions according to the 'city as system' theory. One strategy is to increase resource efficiency of the city system by means, such as enhancing motor vehicle engine technology or fuel to reduce resource use or waste and emissions. For example, electrical vehicles can reduce petrol consumption and, thus, reduce GHG emissions. The other strategy is managing or reorganizing the interaction between various urban sectors to minimise resource consumption, waste or emissions under a given sector's technical conditions. For example, rather than motor vehicle efficiency being the focus, access or communication might be improved. The latter strategy involves the strengthening of the urban system using various policies, for example, land use planning, transport planning, increasing density at urban sub-centres, etc. A typical measure is increasing density and land use mix to reduce VMT (vehicle miles travelled) and, thus, reduce energy consumption and GHG emissions. This way of reorganizing a city system aims to optimise urban processes through redistributing activity within the land use and transport systems. It is focused on improving the sustainability and resilience of a city.

Resilience is allied to, but distinguishable from, sustainability. In both cases, meanings are clearer when the context is specific. For example, resilience can mean, in an engineering sense, the ability of a system to return to an equilibrium or steady-state after a disturbance [24]. In this sense, a city's resilience is determined by its recovery from disturbance, its capacity to rebound [5]. A city's vulnerability to natural hazards and disasters depends on both the magnitude of hazards and the city's internal systems. To reduce the vulnerability of a city to hazards is often seen as one of the main goals of building a resilient city.

Ecological resilience is a broader concept and refers to the magnitude of the disturbance that can be absorbed before the system changes its structure [24]. A city system's resilience in this sense is determined by its ability to persist and adapt to a new environment; a city's resilience reflects its ability to remain within given ecological thresholds, either in the existing environment or in the new environment. A variant on this concept is socio-ecological resilience, which focuses on the changing nature of systems over time, with or without an external disturbance, and taking into account social processes [25,26]. Here, changes in resilience reflect the evolution of a city system. As its systems are strengthened, a city has a stronger ability to resist or adapt to new disturbances, for instance, natural disasters [25,27]. In this perspective, urban resilience is conceived of as the ability of a complex social-ecological system to adapt and, when necessary, transform in response to stresses and strains [28]. Resilient city policies help a city to be in a state of evolutionary resilience.

Effective and resilient urban transportation is a combined result of many effective sub-processes: for example, adaptable land use and an adaptable and diverse transport network respond to social processes (changes in income, ethnicity, lifestyle), economic processes (industrial and commercial development, oil price changes, the pricing of parking) and institutional processes (governance, urban planning, transport planning, road pricing). The association between land use and transport has been widely studied (see Handy, [29]; Crane, [30]; Stead and Marshall, [31]; Litman, [32]; Ewing and Cervero [21]). Most empirical studies find that changes in land use can result in changes in travel demand and, thus, induce changes in transport infrastructure systems. Transport infrastructures and traffic characteristics (congestion) affect location accessibility, which is a major factor influencing land use. When time is taken into account, the interactions between land use and transport become more complex. Wegener [33] summarised the land use-transport system as eight subsystems characterised in terms of time: transport networks and land use often have very slow changes; buildings of workplace and housing have slow changes; employment and population caused by economic development have fast change; and goods transport and passenger travel have immediate change.

In the field of transport, many studies have found that land use policies or transport planning do reduce the costs of transport and GHG emissions from transport. For example, Rodier and colleagues [34] found that more intensive or denser land use can yield a reduction of ~10% in US urban transport activity without reducing accessibility. Ewing and colleagues [35] estimated that shifting 60% of new growth to compact patterns would save 85 million metric tons of $CO_2$ annually by 2030 in the USA. A recent study by Grazi and colleagues [36] shows a potential for changes in urban form to reduce average travel distance by 10% (25% when increasing density to its maximum degree), which, in turn, would lead to an 11% (31% under maximum density) reduction in GHG emissions.

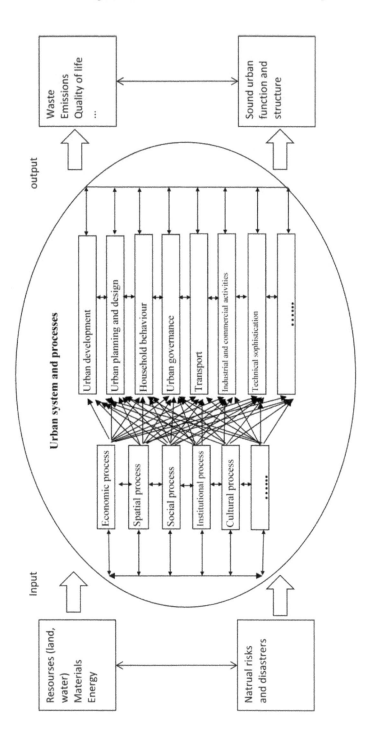

**FIGURE 1:** The urban system and processes.

For environment and planning research, system models have been wide-ly used by researchers to simulate dynamics of the urban system and to evaluate the social and environmental effects of the urban system changes (see a review by [37]). System models are based on systems thinking, which considers individual and separate processes of a city as a connected whole. However, research about urban modelling is still being challenged by a di-versity of methods, metrics, indicators and data. A proper estimation of un-certainty of a system is another challenge to the development of a system model. An urban system and its relationship with natural systems tend to be even more complex and uncertain, because of climate change and related unpredictable natural changes. This calls for new modelling tools and algo-rithms that can take into account phenomena, such as climate change. Re-cently, the topic has been attracting increasing attention from researchers, as the resilience of cities in response to natural disasters and long-term climate change has emerged as a focus of city policies [4,38].

## 1.3 SYSTEMIC APPROACH IN WILUTE

### 1.3.1 CLIMATE ADAPTATION AND MITIGATION IN WELLINGTON

The model is based on Wellington, a fairly compact city core confined by natural topographical features, in the centre of a city region that has sprawled significantly in recent decades. Sea levels in New Zealand rose by 17 cm last century, and they have risen on average 1.7 mm/year over the last 40 years. The city's harbour has experienced an average rise in sea level of about 2 mm per year over the past century. Wellington, like other New Zealand cities, is on the coast and, thus, vulnerable to coastal hazards caused or aggravated by climate change, such as storms and sea-level rise. For example, in Wellington, waves could be 15% higher by 2050 and 30% higher by 2100 [39]. A recent report from the National Institute of Water and Atmospheric Research suggests that Wellington harbour's relative sea level is tracking towards a 0.8 m rise by the 2090s [40], but that for plan-ning purposes, a range of plausible sea-level rise estimates of up to 2.0 m should be considered.

Building resilient cities in New Zealand requires focusing on both mitigation and adaptation. One focus necessitates significant changes to transportation and land use systems in order to reduce carbon emissions. The other focus requires changes to enhance the city's capacity to manage impacts of climate change, such as sea-level rise. In regards to mitigation, New Zealand is on track to meeting its Kyoto Protocol commitment for the period 2008–2012, but has achieved this through afforestation, not emission reduction. In fact, emissions have grown sharply since 1990, from 59.1 million tonnes of carbon dioxide equivalent (Mt $CO_2$-e), to 70.6 Mt $CO_2$-e in 2009, an increase of 19.4%. While agriculture was New Zealand's largest emitting sector in 2009 (32 Mt $CO_2$-e), the growth in emissions is largely attributed to growth in energy emissions, particularly from road transport and electricity generation [41]. New Zealand's road transport emissions increased by 66% over the period, 1990–2009 [42]. Carbon emissions from transport are becoming an increasing concern for the New Zealand community and an embarrassment for the New Zealand government. In addition, traffic accidents and other traffic pollutants, such as NOx, $SO_2$, other toxic waste, water pollution and noise pollution, are contributing factors in local environmental and public heath challenges [43].

### 1.3.2 MAIN PURPOSE OF WILUTE

The objective of the Wellington Integrated Land Use-Transport-Environment Model (WILUTE) is to establish an archetypal projection and assessment system for land use and transport development in the Wellington Region. It is designed as a platform to test and evaluate transport or land-use policies and their interaction, with respect to transport-related environmental and public health effects. It can also be used to assess and forecast the vulnerability of the transport and land use system to sea-level rise. To do this, the model is designed to, firstly, measure current energy consumption and environmental pollutants arising from the transport system and forecast the effects of transport or land use policy options on energy consumption and environmental pollutants from transportation. Secondly, it is designed to assess the public health benefits from transport policies. Public health effects in relation to transport include traffic accidents on roads, pedestrians' and cyclists' exposure to pollutants from road traffic and active travel. According to a report by WHO, transport-related air pollution affects a number of health outcomes, including mortality, nonallergic respiratory morbidity, allergic illness and symptoms (such as asthma), cardiovascular morbidity, cancer, pregnancy, birth outcomes and male fertility. Transport-related air pollution increases the risk of death, particularly from cardiopulmonary causes, and of non-allergic respiratory symptoms and disease.

At the current stage, the WILUTE model is focused on the assessment of the impacts of the transport and land use system on carbon emissions, active travel (cycling and walking) and local residents' exposure to pollutants from road traffic. In the next stage, the WILUTE model will be used to explore other transport-related air pollution impacts, such as health modelling progresses, and to collect health data. Thirdly, the model will be used to predict how the transport system is exposed to sea-level rise and project first-round socioeconomic outcomes of possible policies responding to sea-level rise.

At present, four key questions are being addressed by the model for the Wellington Region:

1.  How does the existing transport and land use system influence carbon emissions and local air quality in the region?
2.  How might future transport infrastructure (e.g., new light rail, new cycle lanes) change current transport mode choices and promote green transportation?
3.  To what extent are current transport and settlements vulnerable to sea-level rise?
4.  How can the capacity of the transport-land use system to respond to sea-level rise be strengthened in future?

The model measures short-term transport activities (e.g., mode choice, route choice, travel time), long-term transport activities (car ownership, travel distance), long-term transport effects caused by socioeconomic activities (e.g., household location and relocation choice and employment location and relocation choice) and the effects of sea-level rise on transport (transport links, passenger traffic), as well as possible transportation results of policies designed to respond to sea-level rise.

The model analyses land use at different scales: buildings, parcels, neighbourhoods and communities, since policies are usually concerned with issues at multiple geographical levels. At the buildings scale, the model uses information on individual properties, such as location, land area, floor area, age, use, site cover, etc. Parcel data, which includes information on boundary, size, land use and subdivision, are used at the parcels level. At the neighbourhood or community level, the model uses information on local facilities and infrastructure. These scales are interconnected in the analysis at the neighbourhood or community level. For example, information on land use at a community level is aggregated from information on individual parcels, which are, in turn, aggregated from individual building data.

WILUTE addresses four main aspects of urban sustainability: economic sustainability; social sustainability; environmental sustainability; and system sustainability. In the modelling process, WILUTE generates a number of indicators of urban sustainability from the perspective of the land use and transport system. The indicators cover the main aspects of urban sustainability (Figure 2). The indicators of travel costs in time and money and population and employment growth measure economic sustainability. The social sustainability indicators include housing affordability, which is indicated by housing price and the supply of houses in terms of types and locations, the factors influencing the risk of traffic accidents (traffic speed and volumes) and the percentage of walking and cycling. The environmental sustainability indicators include air pollution, energy consumption, $CO_2$ emissions, etc., and people's exposure to sea-level rise across different income and ethnic groups (environmental equity) and, particularly, their exposure in terms of residential location.

System sustainability is indicated in a stylised way by its financial capacity and the costs (time, resources and social costs) needed by the transport and land use system to recover to a "normal" situation in the event of a natural disaster associated with sea-level rise. These costs include the costs of relocating residents, industries and facilities. The costs also include the investment in new infrastructure to reduce the impacts of sea-level rise, for example, sea walls and new elevated highways in the most vulnerable areas. The assumption is made that these measures indicate the broad magnitude of cost for a likely response strategy; it is acknowledged that other response strategies are possible.

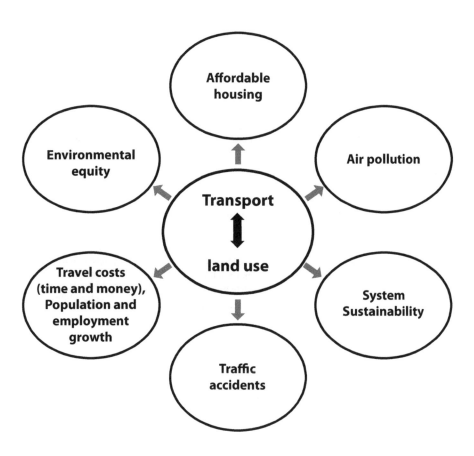

**FIGURE 2:** Urban sustainability indicators.

Table 1 shows how these indicators cover both the human and natural systems in a city. The indicators measure two-way interactions between human and natural systems in a city, as emphasised in writing on social-ecological resilience. The ability of a city system to reduce energy consumption from transport and buildings affects natural systems. Conversely, the vulnerability of the transport and land use system to natural hazards, such as sea-level rise, shows the impact of the natural environment on the city.

**TABLE 1:** Human and natural system interactions and resilience indicators.

| Indicators | | Natural system | | | |
|---|---|---|---|---|---|
| | | Air | Land | Environment and other resources | Natural disasters and hazards |
| Human system | Transport | Air pollution | | | City resilience (ability to reduce energy use and emissions; vulnerability of transport and land use to sea-level rise; costs related to reducing the vulnerability) |
| | Housing | | Housing affordability | Environmental equity related to residential location | |
| | Economic growth | | Population and employment growth | | |

### 1.3.3 SYSTEMIC METHODOLOGY IN WILUTE

As noted above, city system theory is applied in the WILUTE model. The model treats land use, transport and the environment in an integrated way. The model attempts to take full account of the complex interactions and synergies that occur between urban processes (economic, social and spatial process), including household location choice, firm location choice, transportation choices and land use decisions. In the model, environmental factors (e.g., energy use) are treated as endogenous elements in the transportation distribution and mode choice. The environmental effects of land use-transport polices are measured at different levels, including areas, links and sites.

**FIGURE 3:** Architecture of the Wellington Integrated Land Use-Transport-Environment Model (WILUTE).

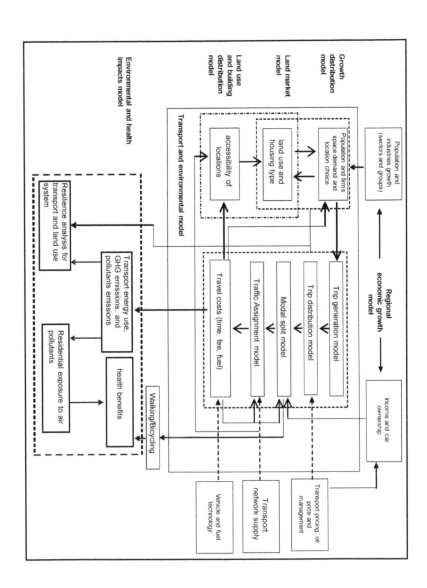

The core of the WILUTE model is derived from the IELT model [44]. IELT refers to an integrated economy, land use and transport system model. It can be used to forecast regional economic growth and changes in land use and transport. An IELT model has already been validated using data from Beijing. The WILUTE model extends the IELT model in three ways. First, land use is modelled in a more precise way, at a parcel level based on individual properties. Secondly, a health impact sub-model is added. Thirdly, a resilience analysis is included. The architecture of the model consists of six sub-models: a regional economic growth model, growth distribution model, land market model, land use and building distribution model, transport and environmental model and environmental and health impacts model (Figure 3).

The regional economic growth model forecasts the growth (or decline) of firms by sector, population by group and the increase of household incomes and car ownership. An input-output function is applied. The growth distribution model distributes the growth (or decline) of population to the local level (land parcel) across the whole region. A multinomial logit (MNL) function is applied for household location choice and employment location choice. The land market model transfers the economic growth into land demand and estimates the land price according to land demand and supply. A dynamic market equilibrium rule is applied in the land market model. The land use and building distribution model distributes space and housing demand to local levels. It is also based on a discrete choice model. The transport and environmental model is derived from the traditional 'four-step' transport demand model and a transport energy and carbon dioxide emissions estimation. In the model, travel costs are transferred into area-based accessibilities, which are major factors in the distribution of economic growth and the land market. The air quality and health benefits model utilizes the estimates of trips and transport energy consumption to measure the transport links' emissions. Transport energy consumption measurement takes into account energy intensity, calculated in litres per 100 km (L/100 km), by different vehicles at different speeds. GHG emissions in transport links are measured from energy consumed in the links. Emission factors are used to quantify GHG emissions per litre of energy consumption. To allow calculation, vehicles are classified in terms of their emission level, such as Euro 3 or 4. The transport links' emissions are transferred into site- and area-based ambient air quality. Public health impacts are simulated by changes in active travel trips (walking and bicycling) and concentrations of air pollutants.

The resilience analysis aims to evaluate how resilient a city's transport and land use system is. WILUTE uses three types of indicators to measure city resilience. The first is a city's capacity to reduce energy consumption from urban transportation in particular. The second is the exposure of the land use and transport system in a city to natural disasters, sea-level rise, in particular. This is measured as the vulnerability of residents, traffic links and traffic flows to sea-level rise. In the analysis, local topography, weather and infrastructure conditions (e.g., flood-proof dikes) are taken into account. The third indicator is the costs related to reducing exposure of the land use and transport system to potential nature disasters to an acceptable level. These costs include the costs of relocation of residents and economic activities and building new infrastructure to reduce the impacts of natural disasters, such as sea walls and dikes, etc.

## 1.3.4 THE MERITS OF THE SYSTEM APPROACH IN WILUTE

The WILUTE model has several advantages, compared to the current most widely used integrated transport-land use models in the world, for example, LTLUP (The Integrated Transportation and Land Use Package), IRPUD (The Institute of Spatial Planning of the University of Dortmund), LILT (The Integrated Land-use Transport model), MEPLAN (The Marcial Echenique Plan), TRANUS (Transporte y Uso del Suelo), DELTA (The Land-use/Economic Modelling Package), POLIS (Projective Optimization Land-Use Information System), MASTER (Micro-Analytical Simulation of Transport, Employment and Residence), etc. [33]. Firstly, while most of the previous models integrate land use and transport, and some partly integrate economic development with land use and transport, none integrates environmental effects with economic development, land use and transport. The WILUTE model integrates economic development, land use and the transport system with transportation energy consumption, greenhouse gas emissions, local air quality and public health co-benefits. In particular, the interactions between environmental effects and transport and land use are considered in the

WILUTE model in two ways. One is that exposure to traffic pollutants due to housing location and traffic flows can affect residential location choices, which, in turn, shape new patterns of traffic flows. The other is that residents' consideration of vulnerability of their houses to sea-level rise affects their location choice and traffic flows, which influence new property development and infrastructure investments.

Secondly, in the model, land use and transport is simulated by a discrete choice approach, which is based on utility theory and has advantages in better modelling individuals' behaviour. Land use modelling runs four calculation processes: the residential location choice model, the firm location choice model, the developer's development location choice model and the developer's land use choice model. These calculation processes are based at a census area unit level, which approximates a community neighbourhood, and is aggregated up from individual buildings and, then, from parcels.

Thirdly, as noted above, many empirical studies have found that individual travel behaviour is the key to more sustainable transportation and more important than technical factors and infrastructure supply. The WILUTE model forecasts travel demand in a disaggregated way based on individuals' travel behaviour data. Therefore, WILUTE has the potential to evaluate GHG emission reduction policies more accurately than previous models, which are based on aggregated travel patterns. Fourthly, the model has a transparent architecture, which can be easily understood by policy-makers and the public. Most of the previous models have been criticised, as they have an architecture that is often seen as "black box" to local government officers [33]. During the development of WILUTE, the model was demonstrated to and discussed with City and Greater Wellington Regional Council officers. Some scenarios were presented, and ongoing discussions are proceeding to utilize and further test the model.

Currently, cellular automata (CA) models are not uncommon. Compared with a CA land use model, the WILUTE model has at least three merits: (1) it achieves a greater integration between land use and transportation, as it can estimate traffic flow, pattern and accessibility and involve traffic outcomes in forecasting the changes in housing location or employment location and, thus, land use; (2) it treats land use as a complex process, which is affected not only by transport accessibility, but also by individual households' choices of location and travel mode; and (3) it can directly quantify the environmental effects of transport and land use policies through a group of indicators: land consumption, transport energy use, emissions and local air quality impacts. To the extent that a CA model has a comparative advantage in simulating land use changes, the WILUTE model could integrate a CA model within its sub-model of land use and building distribution.

Land use or transport policy options are accessed as input variables relating to the district plans, the urban design framework and "design upgrades", the urban growth boundary, transport infrastructure provision (e.g., new light rail), public transit service (e.g., service quality), travel demand management (transport pricing, oil price, parking management), vehicle and fuel technology, etc. These options are amenable to modelling with the WILUTE model, which can be used at different geographical scales—from individual building sites, to neighbourhoods, communities and cities.

## 1.3.5 THE OPERATION OF WILUTE

The WILUTE model is organized as a dynamic GIS (Geography Information System)-based operational model (Figure 4, Figure 5). It runs in annual time steps, progressing through the regional economic growth model, growth distribution model, land market model, land use and building distribution model, transport and environmental model and air quality and health benefits model. The transport model simulates travel for an average working day for transport zones for one year. It also estimates a traffic equilibrium incorporating congestion effects. It provides outputs of accessibility to land uses for the model for the subsequent year. Traffic changes and location changes in housing and employment often do not occur simultaneously. Location changes usually lag behind the traffic changes. The model takes this into account. In the model, the interactions from the transport model to land use are simulated less frequently than both land use and transport model changes. A five-year time lag is used to reflect the features of interaction between transport and land use.

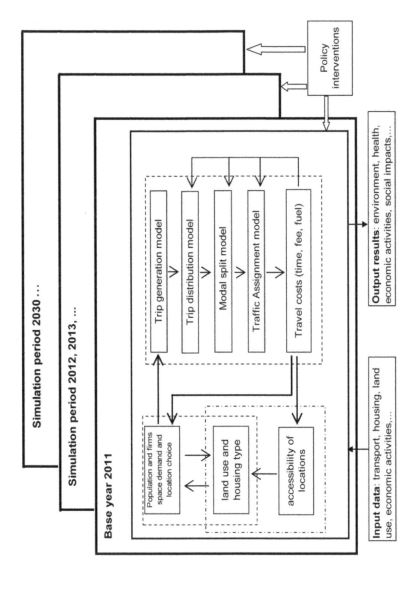

**FIGURE 4:** The operation of WILUTE.

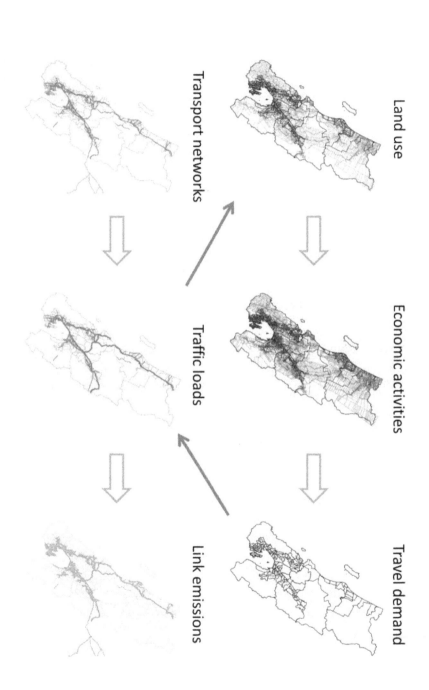

**FIGURE 5:** The GIS data process in WILUTE.

There are two types of input data in the modelling process. One is base year data, including data on transport (traffic survey, transport network, fares, etc.), housing or property, land use, demographics, employment, topographic, etc. The other input comprises policy intervention data, for example, on land use planning, transport planning, road pricing, fuel taxes and behaviour change education—each of which is designed to change personal travel behaviour. The output data show the values of the indicators for urban sustainability that are estimated in the model. The output data also include two important indicators measuring city resilience, as noted earlier. The costs related to reducing vulnerability to climate impacts are not considered in WILUTE at the current stage.

## 1.4 APPLICATION EXAMPLES

### 1.4.1 TRANSPORT RESULTS OF SCENARIOS

Several transport and land use scenarios were composed to reflect transport and environmental effects of different transport-land use policy scenarios, including intensification scenarios, an urban limit scenario, a transit-oriented development scenario, a transport infrastructure scenario and a fuel and traffic pricing scenario. The environmental effects, in particular, energy consumption and carbon emissions, are estimated in each transport-land use scenario. Meanwhile, several sea-level rise scenarios are estimated for each transport-land use scenario. The base year of the model is 2006, and the scenario final year is 2031. This paper focuses on introducing two intensification scenarios to illustrate the model. One is a high intensification scenario; the other is a low intensification scenario.

In the low intensification scenario; the current pattern of low dwelling and population density are assumed to continue at the business as usual level; for new incremental houses; the floor area: land area ratios were changed to the average value measured in 10 minimum traffic zones (at the 2006 business as usual level).

To create the high intensification scenario, four changes to "business as usual" were made to indicate increases of urban intensification. Firstly, for new incremental developments, the land per person for each of the land use types was reduced to the average of the 10 lowest (non-zero) land areas per person for all traffic zones. This proxies intensification policies, such as limits on land supply and a growth boundary. This change also altered the population densities of each traffic zone.

Secondly, we changed the ratio of houses to apartments from 9:1 in base year 2006 to 1:9 (houses: apartments) in 2031 in high traffic zones. This measures the urban infill development policies and policies of building high density apartments in transit areas.

Thirdly, we changed the land per apartment and land per house to the minimum value in 2006 for all traffic zones. The smaller the section, the greater the intensification.

Finally, for individual parcels, the floor area:land area ratios (FARs) were changed from being the average of the 10 lowest FARs to being the average of the 10 highest FARs. This would simulate individual houses or apartments having more base (floor) area for buildings and leave less space for non-building functions, for example, green yards or open spaces.

In the modelling process, individuals were grouped by age and income into five categories: people of aged 0–10 years old; 11–19 years; 20–64 years with no or low income; 20–64 years with medium or high income; and above 64 years old. Each trip was categorized into five groups based on the purpose of the trip, the purpose of the overall journey and the starting place of the trip. These categories were: home-based work trips, for trips that started at home and had the overall journey purpose of going to work; "home-based non-work" for trips that started at home, but were not for the purpose of getting to work; "non-home based trips home", for all trips returning home from any other location; "non-home based trips to work", for trips to work that did not start at the home address; and "non-home trips other", for all trips that did not go to work and did not start at, or return to, the home address.

Figure 6, Figure 7 show the population and transport results of the low intensification scenario and the high intensification scenario, respectively. The modelling results suggest that the high intensification scenario could save about 20,600 trips per day, compared to the low intensification scenario. The exact energy savings and the reduction of social costs of transport are not reported here. However, it is clear that policies designed to enhance urban intensification would reduce the carbon footprint of the city in the future.

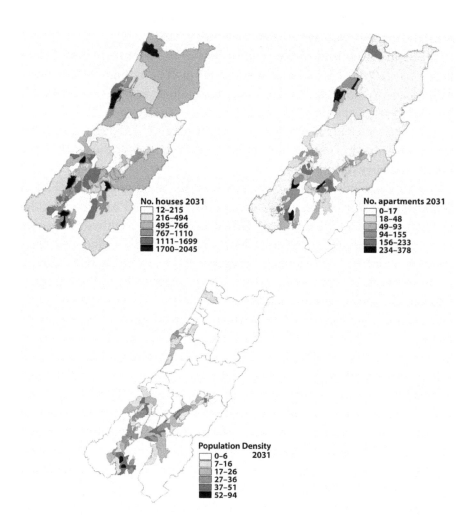

**FIGURE 6:** Population, houses and transport in the low intensification scenario.

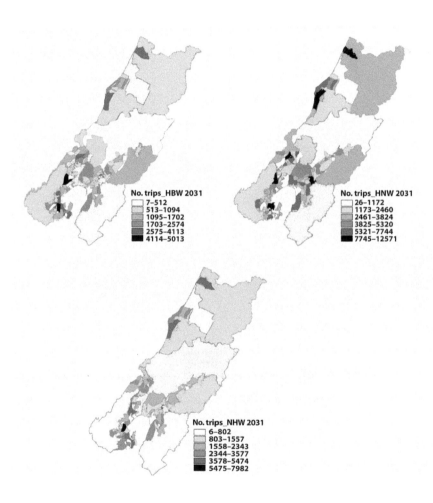

**FIGURE 6:** *Cont.*

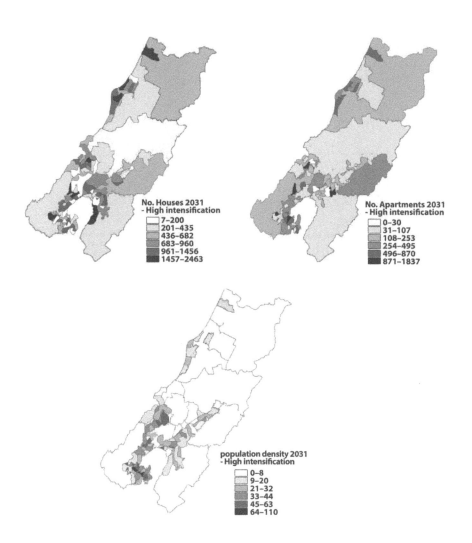

**FIGURE 7:** Population, houses and transport in the high intensification scenario.

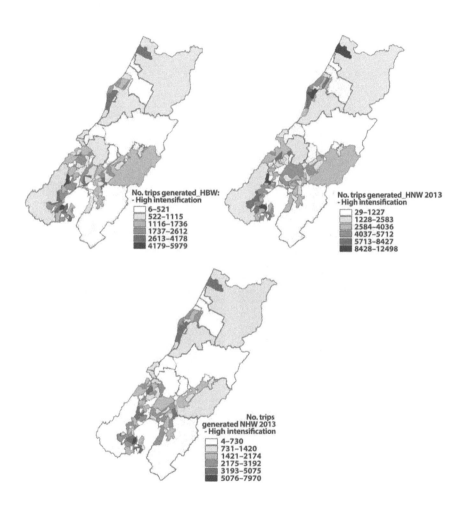

**FIGURE 7:** *Cont.*

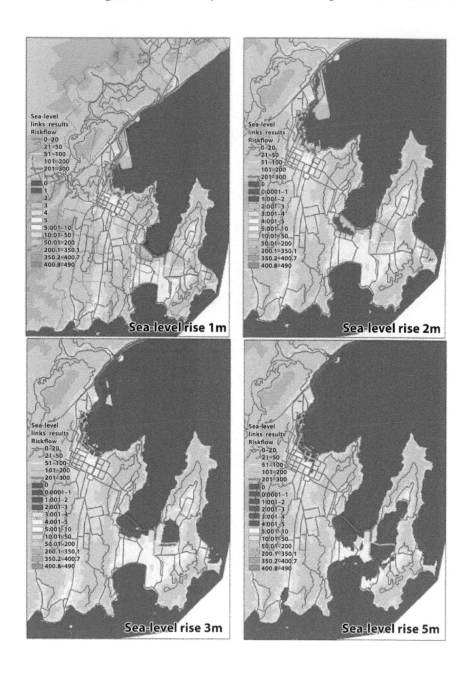

**FIGURE 8:** Vulnerability of traffic links and flows to sea-level rise.

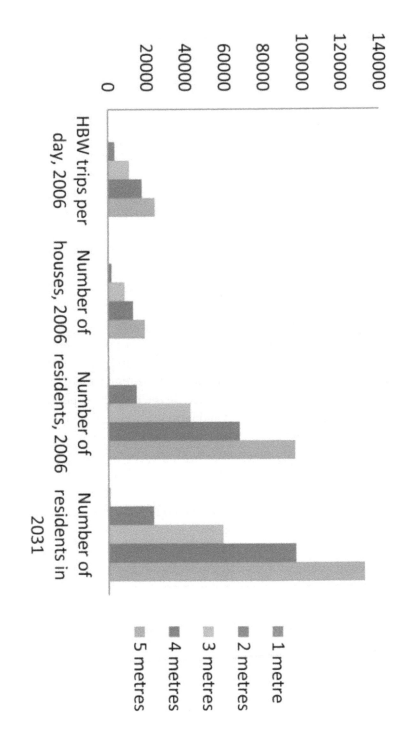

**FIGURE 9:** Vulnerability of people, houses and transport to sea-level rise.

## 1.4.2 TRANSPORT VULNERABILITY TO SEA-LEVEL RISE

The section provides an example of the application of WILUTE for forecasting transport vulnerability to sea-level rise. In the WILUTE model, the vulnerability of people, houses and transport are calculated in different ways. Firstly, vulnerability of people and houses to sea-level rise in an area in a given period (e.g., 2030) is calculated from the distribution of residents, land development and houses through a spatial analysis. The sea-level rise in this given period is measured with consideration of sea-wall and flood protection. Secondly, the vulnerability of transport to sea-level rise is calculated through two indicators. One is the transport network vulnerability to sea-level rise in an area in a given period. It calculates how transport infrastructure is vulnerable to sea-level rise through spatial analysis. The other is the vulnerability of traffic flows by population group. This is calculated from the outcomes of the transport model in WILUTE.

Figure 8 shows the vulnerability of traffic links and flows (standard cars per hour) to sea-level rise, based on transport patterns in 2006. Many major roads in the city are at risk of being flooded, depending on the extent of sea-level rise. We underline that the higher sea-level rise scenarios (above 2 m) are extremely unlikely in the projection period to 2031. Wellington International Airport and its major connection roads would be seriously affected if sea level rise were 3 m. The scenario analysis suggests that nearly 11,000 home-based work trips per day would be affected by a three-metre sea-level rise. Over 42,000 residents would need to be evacuated and 8,000 houses would, in theory, be under the sea if sea level rose by 3 m (2006) (Figure 9). Nearly 60,000 residents would need to be evacuated by 2031 if sea level rose by 3 m.

The effects of sea-level rise may not be equitably distributed between places and between different groups in terms of income or ethnicity, since the transport-land use system has high spatial heterogeneity features. Vulnerability in this context is interpreted specifically as living in a dwelling vulnerable to sea-level rise (it excludes, for example, impacts on people's trips to work or other forms of vulnerability). This underlines the environmental inequity regarding the impacts of climate change and sea-level rise. Figure 10 shows that of those people below 10, on low or no incomes, or over 65 years old, 54% are vulnerable. Only 24% of those on medium to high incomes aged 20–64 are vulnerable. And of those aged 11–19, only 22% are vulnerable.

## 1.5 DISCUSSION AND CONCLUSION

Making cities more resilient has become one of the most important goals of urban sustainability in many countries. How to build a resilient city in order to respond to climate change and other possible disasters, such as earthquakes and tsunamis, is of major interest to planners, politicians and the public. For researchers, great efforts have been made to contribute to resilience policies, but there are still major gaps in the existing literature on urban resilience. In particular, we lack a sound method through which to properly evaluate alternative resilience policies.

Policies for resilient cities are interpreted here as those strengthening the urban system's capacity to change in response to economic and natural shocks to the system [26]. The various human-built systems of a city and the natural system are complex. In particular, the complexity of urban systems presents a major challenge for identifying environmental impacts of urban processes and for evaluating the efficiency of policies in reducing environmental and health impacts. Recently, simulation tools have seen significant developments in relation to transport emissions and other health-related effects. The approaches range from operations research (OR) to system dynamics (SD) and discrete-event or discrete-agent system simulation approaches (DS). This paper has introduced an integrated land use-transport-environment model, which can be used to evaluate city resilience outcomes under different policy scenarios. City resilience is measured by its capacity to reduce energy consumption and GHG emissions, the vulnerability of its transport and land use system to sea-level rise and the costs related to reducing the vulnerability to a safe level by taking into account its financial capacity. The model is designed to estimate the complex links between urban economic activities, household and firms' location choices, transportation and land use, building on accepted theories of urban systems. The design of the model can enhance our existing knowledge of a systemic approach to study urban transport and land use policies and effects on urban sustainability.

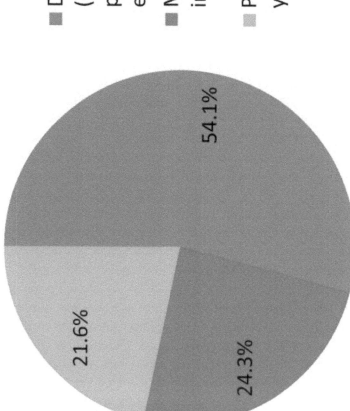

FIGURE 10: Vulnerability of different groups to sea-level rise (5 m).

Future modelling of urban systems still faces challenges, including the application of models in practice. The approach to treating "dynamic" issues within a model needs further attention. Most models, including the WILUTE model, project the state of the urban system and its effects on the environment in the future, relative to a base year. The future year is seen as a user-specified horizon year for the model. The dynamic features of the urban system are treated as the evolution of the system state from one point in time (the base year) to another (e.g., horizon year). Many models assume that the system equilibrium (within land use system, transport system or both) can be achieved by the evolutionary process of the system.

A model is able to simulate the dynamic processes of the system through appealing to the mathematical conditions for equilibrium to represent the system situation at a given point in time. In fact, urban system dynamics will be much more complicated than we can simulate using such processes across different time-frames. For example, in the WILUTE model, it is assumed that there is a five-year time lag between changes in land use and transport, and a five-year time lag is used to reflect such features of dynamics in the transport and land use system. However, some fast-acting dynamics in transport (e.g., departure time, travel path choice) may not be reflected well. Some very slow dynamics (e.g., urban land use patterns) may only be partly included in a model's time scale. Combining various dynamic processes within one overall model is still a big challenge to a system approach when it is applied to evaluate urban sustainability.

Secondly, a proper estimation of the uncertainties in any urban system is another challenge to the development of such models. Uncertain interactions between urban processes and between human activities and environmental processes are typical features of the urban system. Uncertainty of an urban system is an on-going topic of discussion of integrated modelling. How to enhance the capacity of a model to respond to uncertainties in urban systems has become a new challenge to systemic modelling. There are various possible ways to approach this. One is to increase the comprehensiveness of the policy alternatives or scenarios. Another is to increase the flexibility of the model regarding the calculation process and data needed.

When it comes to policy practice, individual policies designed to build a resilient city or enhance a city's resilience need to be integrated; overall resilience is more than the sum of the parts. In the transportation field, transport infrastructure projects, land use planning, housing policies and travel demand policies should be integrated in order to reduce transport-related energy consumption and GHG emissions. Different aspects of city sustainability should be included in resilient city policies. In particular, environmental justice across different income and ethnic groups is often ignored. As Newman and colleagues [4] argued, a resilient city will reduce its ecological footprint, while simultaneously improving the quality of life it provides to the population. Improving social and environmental equity is also one of the objectives of building a resilient city [45,46]. Integrated solutions, which can be explored by models, such as the WILUTE model, need to be addressed in future policy-making to create resilient cities.

## REFERENCES

1. United Nations, World Urbanization Prospects, the 2009 Revision: Highlights; Department of Economic and Social Affairs, Population Division: New York, NY, USA, 2010.
2. IPCC (Intergovernmental Panel on Climate Change), Climate Change 2001: The Scientific Basis; Cambridge University Press: Cambridge, UK, 2001.
3. Pielke, R.A., Sr.; Marlans, G.; Betts, R.A.; Chase, T.N.; Eastman, J.L.; Niles, J.O. The influence of land-use change and landscape dynamics on the climate system: Relevance to climate-change policy beyond the radiative effects of greenhouse gases. Phil. Trans. R. Soc. Lond. A 2002, 360, 1–15.
4. Newman, P.; Beatley, T.; Boyer, H. Resilient Cities: Responding to Peak Oil and Climate Change; Island Press: Washington, DC, USA, 2009.
5. Vale, B.; Campanella, T.J. The Resilient City: How Modern Cities Recover from Disaster; Oxford University Press: New York, NY, USA, 2005.
6. IPCC (Intergovernmental Panel on Climate Change), Fourth Assessment Report: Climate Change; Cambridge University Press: Cambridge, UK, 2007.
7. Rahmstorf, S.; Foster, G.; Cazenave, A. Comparing climate projections to observations up to 2011. Environ. Res. Lett. 2012, 7, 1–5.
8. UN-HABITAT. State of the World's Cities Report 2008/2009: Harmonious Cities. Available online: http://www.unhabitat.org/pmss/listItemDetails.aspx?publicationID=2562 (accessed on 12 March 2013).
9. World Health Organization, Health in the Green Economy: Health Co-Benefits of Climate Change Mitigation-Transport Sector; World Health Organization: Geneva, Switzerland, 2011.
10. IEA, CO2 Emissions from Fuel Combustion Highlights; International Energy Agency: Paris, France, 2012.
11. IEA, World Energy Outlook; International Energy Agency: Paris, France, 2006.

12. IEA, Energy Technologies for a Sustainable Future: Transport; International Energy Agency: Paris, France, 2004.
13. Puentes, R.; Tomer, A. The Road Less Traveled: An Analysis of Vehicle Miles Traveled Trends in the U.S.; Brookings Institution: Washington, DC, USA, 2008.
14. Newman, P.; Kenworhy, J.R. Peak Car Use: Understanding the Demise of Automobile Dependence. WTPP 2011, 17, 31–42.
15. Ekins, P. How large a carbon tax is justified by secondary benefits of the CO2 abatement? Resour. Energy Econo. 1996, 18, 161–187.
16. World Health Organization, The Global Burden of Disease: 2004; World Health Organization: Geneva, Switzerland, 2008.
17. World Health Organization, Global Health Risks: Mortality and Burden of Disease Attributable to Selected Major Risks; World Health Organization: Geneva, Switzerland, 2009.
18. Banister, D., Button, K., Eds.; Transport, the Environment, and Sustainable Development; Spon Press: London, UK, 1993.
19. Batty, M. Cities and Complexity: Understanding Cities with Cellular Automata, Agent-Based Models, and Fractals; MIT Press: Cambridge, MA, USA, 2005.
20. Forrester, J.W. Urban Dynamics; MIT Press: Cambridge, MA, USA, 1969.
21. Ewing, R.; Cervero, R. Travel and the Built Environment—A Meta-Analysis. J. Am. Plann. Assoc. 2010, 76, 265–294.
22. Berkes, F.; Folke, C. Linking Social and Ecological Systems: Management Practices and Social Mechanisms for Building Resilience; Cambridge University Press: Cambridge, UK, 1998.
23. WHO, World Health Report 2002: Reducing Risk, Promoting Healthy Life; World Health Organization: Geneva, Switzerland, 2002.
24. Holling, C.S. Engineering resileince versus ecological resilience. In Engineering Within Ecological Constraints; Schulze, P.C., Ed.; National Academy Press: Washington, DC, USA, 1996; pp. 31–44.
25. Scheffer, M. Critical Transitions in Nature and Society; Princeton University Press: Princeton, NJ, USA, 2009.
26. Folke, C.; Carpenter, S.T.; Walker, B.; Scheffer, M.; Chapin, T.; Rockström, J. Resilience thinking: Integrating resilience, adaptability and transformability. Ecol. Soc. 2010, 15, 20–28.
27. Nelson, D.R.; Adger, W.N.; Brown, K. Adaptation to Environmental Change: Contributions of a Resilience Framework. Annu. Rev. Env. Resour. 2007, 32, 395–419.
28. Carpenter, S.R.; Westley, F.; Turner, G. Surrogates for resilience of social-ecological systems. Ecosystems 2005, 8, 941–944.
29. Handy, S. Critical Assessment of the Literature on the Relationships Among Transportation, Land Use, and Physical Activity; TRB Special Report 2821996; Transportation Research Board: Washington, DC, USA, 2005.
30. Crane, R. The Influence of Urban Form on Travel: An Interpretive Review. J. Plan. Lit. 2000, 15, 3–23.
31. Stead, D.; Marshall, S. The Relationships between Urban Form and Travel Patterns: An International Review and Evaluation. EJTIR 2001, 1, 113–141.
32. Litman, T. Land Use Impacts on Transport:: How Land Use Factors Affect Travel Behavior; Victoria Transport Policy Institute: Victoria, BC, USA, 2012.

33. Wegener, M. Overview of land-use transport models. In Handbook of Transport Geography and Spatial Systems; Hensher, D.A., Button, K.J., Eds.; Pergamon/Elsevier Science: Kidlington, UK, 2004; Volume 5, pp. 127–146.
34. Rodier, C.J.; Johnston, R.A.; Abraham, J.E. Heuristic policy analysis of regional land use, transit and travel pricing scenarios using two urban models. Transport. Res. D 2002, 7, 243–254.
35. Ewing, R.; Bartholomew, K.; Winkelman, S.; Walters, J.; Chen, D. Growing Cooler: The Evidence on Urban Development and Climate Change; Urban Land Institute: Washington, DC, USA, 2007.
36. Grazi, F.; van den Bergh, J.C.J.; van Ommeren, J. An Empirical Analysis of Urban Form, Transport, and Global Warming. Energ. J. 2008, 29, 97–122.
37. Albeverio, S., Andrey, D., Giordano, P., Vancheri, A., Eds.; The Dynamics of Complex Urban Systems: An Interdisciplinary Approach; Physica-Verlag Heidelberg: New York, NY, USA, 2008.
38. Hickman, R.; Ashiru, O.; Banister, D. Transport and climate change: Simulating the options for carbon reduction in London. Transp. Policy 2010, 17, 110–125.
39. Ministry for the Environment of New Zealand. Adapting to climate change. Available online: http://www.mfe.govt.nz/issues/climate/adaptation/index.html (accessed on 18 Feburay 2013).
40. Bell, R.; Hannah, J. Sea-Level Variability and Trends-Wellington Region: Prepared for GWRC; NIWA: Wellington, New Zealand, 2012.
41. The Ministry of Environment. New Zealand's Greenhouse Gas Inventory 1990–2009, Environmental Snapshot; The Ministry of Environment: Wellington, New Zealand, 2011.
42. The Ministry of Environment. New Zealand Climate Change; The Ministry of Environment: Wellington, New Zealand, 2006.
43. The Ministry of Environment. Health Effects due to Motor Vehicle Air Pollution in New Zealand; The Ministry of Environment: Wellington, New Zealand, 2002.
44. Zhao, P. Urban Form and Transport Energy Use in Beijing: A Long-Term Prospective Analysis; Peking University: Beijing, China, 2009.
45. Boone, C.G. Environmental justice, sustainability and vulnerability. Int. J. Urban Sustain. Dev. 2010, 2, 135–140.
46. Pierce, J.C.; Budd, W.W.; Lovrich, N.P., Jr. Resilience and sustainability in US urban areas. Environ. Polit. 2011, 20, 566–584.

# CHAPTER 2

# Towards Sustainable Cities: Extending Resilience with Insights from Vulnerability and Transition Theory

LEANNE SEELIGER AND IVAN TUROK

## 2.1 INTRODUCTION

Cities at all stages of development are exposed to increasing economic and environmental pressures and instabilities associated with globalization, urbanization, climate change and resource depletion. A United Nations study in 2011 found that 60 percent of the people living in the world's 450 largest cities were at a high risk of exposure to at least one natural hazard [1]. Most of these are in South-East Asia, with a few in North and South America, and surprisingly few in Europe and Africa. The specific nature of these shocks and stresses are varied ranging from cyclones, flooding to drought and landslides. The top five mega-cities at high risk to one hazard are Tokyo, Delhi, Mexico City, New York and Shanghai.

These hazards are often difficult to anticipate and may interact in ways that amplify their consequences. For example, in New York, where there is a medium risk of cyclones and high risk of floods, Hurricane Sandy left millions of residents without power, closed one of the world's main stock exchanges and cost tens of billions of dollars in property damage and lost business. Despite large parts of the city being inundated the basic infrastructure was soon restored, the economy rebounded, and health care and social security systems protected most households whose lives were temporarily disrupted. New York's ability to rebound so quickly was partly due to the city's capital resources as well as financial and social networks.

Not all cities would be able to recover this quickly. The global financial crisis has in many places ushered in a period of austerity, heightening the risks of adversity for urban citizens and governments, making it difficult for them to cope with these kinds of unexpected events. The capabilities of cities to cope with hazardous occurrences and adapt to unfavorable conditions is crucial to their prospects of sustained growth and development. The concept of resilience has become popular in referring to the essential attributes of cities that enable them to deal with disasters and other threats over which they have little control.

Africa, despite it falling outside the prime high risk zone for exposure to natural hazards, cannot afford to be complacent. Urban population growth is expected to be mostly a developing world phenomenon with the continent expecting a 0.9 billion increase in its population [1]. For example, Juba, the capital of the new Republic of Sudan is one of the fastest growing cities in human history, because of the large numbers of people displaced from the surrounding region by rural famine and civil conflict [2]. It is devoid of formal institutions and infrastructure and residents are in dire need of improved living standards and security. More than half of its population is living below the poverty line and public spending is dependent on foreign aid and intermittent oil revenues.

What makes cities robust and resourceful in the face of such diverse environmental, social and economic difficulties? Their ability to maintain vital functions while at the same time adapting and developing in the light of changing circumstances is crucial for successful urban performance into the future. Sandy is an example of an increasingly frequent extreme event that brought one of the world's most prosperous cities to a standstill, albeit temporarily because it was well prepared and protected. Juba lacks the resources to cope with unprecedented population growth that exposes it to overcrowding, misery and disease in the absence of essential services. This paper discusses some of the basic requirements for cities in different circumstances to respond effectively to both sudden and slower-moving disturbances. It draws on the frameworks of resilience, vulnerability and transition to identify some of the inherent attributes that cities require for sustained growth and development.

The concept of resilience has gained rapid acceptance as a way to describe key features of durable ecological, social and economic systems. It seeks to explain how such systems withstand, recover from and reorganize in response to turbulent conditions. There are three main interpretations of resilience, which shed light on different kinds of response to crises. They emphasize the need to 'bounce back', 'bounce forward' and evolve in dynamic situations. However, this paper argues that resilience could be further enhanced by a sharper focus on human vulnerability and innovation. We turn to theories of vulnerability and transition to show how they incorporate these aspects more fully. Vulnerability pays particular attention to the specific groups of people that are most susceptible to harmful effects. Transition theory draws attention to the role of technology in enabling change and development.

This paper draws on examples of small, medium and larger cities to illustrate the breadth and depth of the concept of resilience and its application. The focus in these examples is on how change, both rapid and incremental, affects the complex interactions within and between the ecological and social aspects of cities, highlighting specifically the economic dimension of the latter.

Engineering resilience cannot be applied only to well-resourced cities in advanced economies. Nor can one surmise that the transformation implied by socio-ecological resilience is limited to developing countries. The kind of resilience that is required depends on the nature of the change and its expected impact on a city's resources and people. Similarly, the type of resilience is not fixed to a particular kind of concern like environmental, social or ecological. There are times when multi-equilibria resilience is just as applicable to concerns about ecological limits as it is to concerns about market stability. The examples discussed in the paper are presented to show that the type of resilience required is always dependent on the time frame and the location of the issue at hand. Flexibility, stability and innovation might mean different things in different places but they are essential capabilities that are critical for building sustainable cities.

## 2.2 FRAMING RESILIENCE WITHIN THE SOCIAL SCIENCES

The concept of resilience came to prominence with the insights generated by Holling in the discipline of ecology during the 1960s and early 1970s [3]. It was originally used to describe the persistence of natural systems in the face of disturbances and their ability to renew and reorganize themselves [4,5]. It was a technical, ecological term and was used to describe whether a social or physical system was able to absorb and recover from a disturbance. There were no value judgments inherent in describing whether such systems could withstand shocks. It was essentially factual statements: either the system could or could not.

Now, with its greater application in social science, the concept of resilience is also being used normatively to suggest that the ability to recover and survive is a desirable feature. This must be treated with caution because it is important not to assume that sustaining a system is necessarily desirable, especially for social systems, since the existing state of the system may well not be the optimum or 'normal' condition. Greater dynamism and upheaval may improve many systems. Furthermore, there are diverse and often conflicting values within all social systems, so it has to be established whether one outcome is preferable to another, and not taken for granted that the existing situation should be maintained.

There is another reason for being aware of the normative application of the concept of resilience, particularly when considering the interaction between social and ecological systems in cities and other human constructs. Social systems are strongly influenced by human intentions and perceptions, which in turn are guided by different and competing interests. The physical environment of cities is manipulated for human use with particular objectives and functions in mind. They may be to sustain certain kinds of commercial activities, to meet basic household needs, or to support recreational, artistic and cultural functions. These value land and related resources in different ways that are not easily reconciled. Hence the specific vantage point and framing values of those in positions of power and influence over the built environment are all-important. Cities and other socio-ecological systems are not neutral, disinterested arrangements exposed to immutable external forces with inevitable outcomes. They are subject to human agency and socially constructed in a variety of ways that need to be unpacked and elucidated.

Resilience has varied meanings in different disciplines and therefore the interdisciplinary use of the term can create confusion [4]. In biophysical sciences, resilience is traditionally referred to as the biological ability to persist in environments, whereas in economics it is often understood as the economy's ability to return to a steady state, or equilibrium [6].

In this paper, resilience is used to describe the 'social', 'economic' and 'environmental' responses to change in cities and the interactions between them at various scales. The emphasis is on the type of resilience that exists or is sought. It does not restrict a particular type of resilience to a specific discipline. When engineering resilience is discussed for example, the emphasis is on how ecological, economic or social systems can return to a previous state. In multi-equilibria resilience, the focus is on adapting these systems to maintain stability. On the other hand, when socio-ecological resilience is being investigated the dynamic nature of resilience is emphasized. Resilience is therefore not limited to persistence or restoration in ecology but could include adaptation. Similarly, it is not restricted to the maintenance of equilibrium in economics but could entertain the structural transformation of the economy.

Ambiguity and vagueness can be minimized when using the concept in this interdisciplinary fashion by carefully specifying the term each time so that its scope and meaning are clear. The context in which it is being used also needs careful definition so that there is little doubt about what is being discussed.

One requirement is to define the geographical and temporal boundaries under discussion—what scale of territory is relevant and over what time period [4,7]. Just how resilient New Orleans was against Hurricane Katrina depends on how the area is defined, for example, whether the affluent suburbs on higher ground are included with the poorer, low-lying areas. Assessing New York's resilience to Sandy depends on whether one is judging its recovery in days, months or years. In practice, there are difficult judgments to be made in defining the relevant system and to avoid excluding critical elements without good reason.

Second, one needs to recognize that different challenges necessitate different approaches to measuring resilience [7]. A sudden shock that causes violent disruption by breaching a threshold is quite different to gradual and persistent pressure for change. Monitoring the resilience of an area hit by an extreme climatic event is very different to assessing the slower processes of economic or social restructuring. Both can have pervasive detrimental consequences. One is external and immediate, whereas the other is ongoing and more internalized. Juba may appear to be just about surviving the effects of rapid urbanization on a day-to-day basis by avoiding overt social disorder and unrest, but the risks of damaging human outcomes need to be judged over a period of years as pressures accumulate and the capacity of local systems to cope is tested to the limit.

Third, there is a useful distinction to be made between specific and general resilience [8,9]. Specific resilience refers to the response of a system to a particular kind of change. For example, it could refer to how a coastal city adapts to rising sea levels caused by global warming. General resilience refers to a system's capacity to respond to a range of risks and uncertainties. The relationship between specific and general resilience is significant because an increase in specific resilience can reduce general resilience if it constrains flexibility to cope with other changes [10], perhaps by narrowing the options available through tying it into a particular path. Fixed capital investment in roads and other resource-intensive infrastructure is a classic example for cities. This can limit their capacity to absorb rising energy prices and reconfigure how they function in response to growing resource scarcity and pressure to raise efficiency.

Besides the framing issue, there are two other considerations to take into account when translating resilience from a natural to a social context, namely determinism and power. First, the ecological sciences tend to be more deterministic than the social sciences, with greater confidence about the possibility of identifying system boundaries, thresholds and crisis points [11]. In cities and similar systems, these are probabilities or tendencies rather than foregone conclusions. Within the social world it is much more difficult to understand and anticipate the way systems develop and change. The door often needs to be left open for human ingenuity, technology and collective determination to intervene and avert a looming disaster or shift the trajectory out of its difficulties. Civic society may do what is required to regenerate a degraded ecosystem or revitalize a local economy under threat. A crisis can galvanize exceptional public effort and support, at least temporarily.

Second, resilience is not a neutral attribute since it may have contrasting and contested implications, prompting questions about who benefits from resilience [11,12]. Some sectional interests may gain from the process of building resilience and others may lose. Although cities are shared spaces with common destinies to some extent, they are also places from which particular segments of the population may be excluded, whether deliberately or unintentionally, and through physical or financial means. Hence values associated with justice and fairness may loom large in strategies to promote urban resilience, both in terms of how decisions are made and how the burdens and benefits are distributed.

Different kinds of resilience may be required at different times in the history of a city. Should cities like New York restore the same subway system that was flooded after Sandy or should they redesign and improve aspects of it in the light of altered climatic conditions and the availability of superior technologies? A disaster can be a catalyst for radical reform by exposing the limitations of existing arrangements and questioning the wisdom of prevailing practices [13]. There are three main interpretations of resilience in the literature: engineering (bounce-back), multi-equilibria (bounce-forward) and socio-ecological (evolutionary). These are best thought of as different responses to external shocks and stresses. In the following section, we discuss the insights from each of them within the context of sustainable cities.

## 2.3 WHAT KIND OF RESILIENCE IS REQUIRED FOR SUSTAINABLE CITIES?

### 2.3.1 ENGINEERING RESILIENCE

This is by far the most common meaning of resilience in the popular discourse and in government policy. In practical applications, it is most widely used in disaster or risk management. The focus is on whether a city or other system can recover its population, infrastructure and institutions following a catastrophic event (for recent examples, see [14,15]). The question is how fast and efficiently the system returns to a steady state. The emphasis is on resisting disturbance and conserving what exists. In natural resource management this leads to an approach that tries to optimize and control the flow of resources [16].

Several examples of this can be found in the United Nations Office for Disaster Risk Reduction campaign to make cities resilient [17]. Participating cities are encouraged to be pro-active in allocating sufficient budget for disaster relief, mobilizing citizens, building capacity in emergency services and putting infrastructural measures in place to reduce risk.

The Austrian city of Leinz, situated in the Eastern Alps, is one of their model cities. Plagued over the centuries by recurrent devastating floods that have destroyed buildings, bridges and other infrastructure, the municipality developed a specialized department to deal jointly with environmental and disaster management. This department co-ordinates all the volunteer emergency services and maintains the ecological functions of the river. Water gauges and retention basin monitoring are linked to authorities. If critical levels are reached, authorities are alerted and a siren warns the public.

The concept of resilience has also been used by social scientists to describe the strength of communities in the face of adversity. Social resilience in an engineering sense is the ability of groups to deal with external pressures without affecting their stability and cohesion [18]. For example, it could mean being able to absorb rising immigration flows without provoking conflict and disorder. London and Amsterdam are often held up as examples of relatively tolerant, cohesive cities. Social resilience could also mean being able to adapt systems (multi-equilibria resilience). In Fortaleza-Cearà Brazil, a slum community started their own bank to fund urban agriculture and other community needs when they could not access established bank finance [19].

Engineering resilience can be linked to the use of equilibrium in mainstream economics. A shock or disturbance may move an economy off its stable state—perhaps into recession and higher unemployment—while self-correcting market forces should in theory bring it back [20]. The resilence of different economies is assessed by their susceptibility to being diverted from their established paths and their response times to recover to a position where labor and other resources are more or less fully employed. One implication of this perspective is that resilient economies do not change their basic structure or function over time. The emphasis in this notion of resilience is on maintaining stability and the persistence of core functions.

Various mechanisms are built into the economy to ensure that the system is not thrown off balance [21]. The role of prices in regulating changes in the demand and supply of resources is an obvious example. Cities consist of all kinds of markets—for labor, land, housing and other types of property—in which these market interactions are commonplace. Yet, urban economies can still be forced off their established trajectory by many external factors. It could be pressures of intensified global competition, changes in consumer taste or demand for local products, or technological breakthroughs achieved by rivals. How cities respond depends on accumulated strengths in their institutions, infrastructure and other assets. Their economies may return to their previous position through some adjustment in wages, property prices or a more creative response, thereby demonstrating bounce-back resilience, or they could fail to cope effectively and stagnate or enter an era of decline.

In New York, it took about a week after Sandy for the electricity to be restored and the main subway lines to be operational. This called for an engineering approach to resilience, based on stability within the core infrastructure networks for the basic metabolism of the city to work. Without these essential services up and running, large parts of the economy could not have functioned, resulting in a loss of output and income. In a robust and resourceful urban economy, this type of resilience is possible and desirable in many respects, as the city operates relatively efficiently in the first place. Of course there may be other features of the city associated with social inequality and environmental degradation that are far less satisfactory.

Applying engineering resilience to places like Juba where most of the city lacks basic infrastructure and other formal systems of social, economic and environmental protection would not be very meaningful. There is effectively no established economy outside the government sector and rising levels of unemployment and poverty. Applying engineering resilience within this context would not offer a great deal. Instead, there is a strong case for some form of transformation and development to improve conditions all round, without degrading the natural ecosystems or informal social systems on which the city also depends.

In every city there are bound to be some parts of the fixed infrastructure or established institutions that do not warrant a return to the status quo and are not worth preserving. If it is feasible and affordable, change may be better than sustaining or coping with unsatisfactory systems. In New York, smart-grid technology would enable outages to be identified, isolated and repaired more quickly, and buried cables would be less exposed during storms. Similarly, with rising sea levels and more frequent and severe storms, steps might be taken to reconfigure parts of the shoreline or install sea barriers to reduce flooding. The alteration or transformation of existing systems and practices has given rise to other interpretations of resilience.

## 2.3.2 MULTI-EQUILIBRIA RESILIENCE

This concept of resilience emerged within ecology out of a recognition that disturbed systems did not always return to the same steady state. While many features might look similar, it was a somewhat different system or regime that re-emerged after a shock [16]. Accordingly, the possibility of more than one outcome or state of equilibrium is accepted. Unlike engineering resilience, multi-equilibria resilience regards systems as composed of many equilibria and able to shift from one regime to another. The transition to a different outcome depends on the system being disrupted by reaching a critical threshold.

Whereas engineering resilience focuses on the efficiency with which a system can recover, multi-equilibria resilience focuses on the robustness of the system, i.e. how long it can remain in a particular state and withstand change before reaching a tipping point and moving to a new regime. While the idea of a single equilibrium is rejected, it remains similar to engineering resilience in that its key feature remains stability rather than progressive change [11]. While this understanding of multiple equilibria was initially resisted by some ecologists, it is now widely accepted [16].

Adaptation is a core feature of an equilibrium approach to resilience. Unlike engineering resilience that seeks to restore a system affected by change, multi-equilibria resilience seeks to adapt the system to better cope or eliminate the stress or shock at hand. Various kinds of socio-ecological adaptations are engineered within cities to prevent pollution levels from tipping critical life-sustaining environmental limits or creating unhealthy dependencies. In Lagos, Nigeria, the government developed a bus rapid transit (BRT) system to reduce congestion and air pollution [22]. This helped to reduce carbon dioxide emissions by 13 per cent and travelling time for passengers, making the city more livable. In Singapore, to avoid dependency on piped water from Malaysia, the city constructed reservoirs for rainwater, developed water treatment plants to re-use water and re-paired existing leaks to adapt to reduce water consumption.

A multi-equilibria perspective on resilient cities is also optimistic in envisaging alternative stable positions that can rectify some of the stresses under a current regime. Pendall [7] uses the example of Hurricane Katrina, which left hospitals, transportation systems and neighborhoods in disarray in New Orleans. Yet, it also exposed the pre-disaster situation as an unacceptable target for recovery. People did not want to return to the pre-existing situation characterized by overcrowding and inequality in services. There was compelling pressure for a new improved regime.

The prospect of alternative equilibria within an economy also raises the possibility that it may not be functioning optimally. Markets tend to be driven by self-interest and short-term considerations rather than the long-term horizons inherent in sustainability thinking. This became all too apparent with the global financial crisis, when concerns about the present completely obscured the future. For similar reasons, city planners generally struggle to get property developers and investors to take a longer-term and broader view of the neighborhoods, retail parks and business precincts they create. Consequently, the objectives of quality, livability, distinctiveness, public transport accessibility and resource efficiency get relegated below immediate commercial imperatives [20]. The outcome is often poorly-designed and badly-integrated developments with detrimental environmental and social effects.

A rather different illustration of the problem of convenient compromises relates to the close relationships that often exist between city governments and the providers of key utilities. These are a source of stability, but tend to prevent new suppliers from entering expanding markets for energy, telecommunications, water and waste collection. This often inhibits the introduction of newer, cleaner and more cost-effective technologies, thereby inflating the economy's cost base and restricting employment creation. A more open and enlightened approach might help to shift conditions towards a more advantageous economic and environmental position.

The idea of multiple equilibria captures the shifting composition of city economies as a result of wider competitive forces and developments in technology. Cities have always had to adjust or adapt their structures to changing conditions and 're-invent' their purpose and identity in the face of shifts in the economic environment or the loss of comparative advantage they previously enjoyed [20,23]. De-industrialization has been a particular challenge to the first generation of large cities that emerged with the growth of manufacturing. It has required the development of new knowledge-based, creative and service industries, the reorientation of local educational institutions and the reshaping of property markets [24]. Cities founded upon single natural resources or specialized industries have often faced the biggest difficulties of diversification and repositioning.

Nairobi (Kenya) is an example of a city in the process of recovering from deindustrialization induced by structural adjustment policies through developing new markets and new technologies. Key parts of government and the private sector are actively pursuing the economic potential of mobile phones and related financial services with the vision of the city becoming a technology and innovation hub [25]. Substantial investment by the Kenyan government in undersea internet cables has been instrumental in boosting bandwidth and cutting prices of internet access. Innovation in cashless payment systems has also stimulated many new businesses engaged in developing all sorts of mobile applications. The start-up and growth of telecoms-based enterprises has been reinforced by increasing amounts of domestic and international investment in business incubators. Major global companies such as Google, Nokia, Microsoft, IBM, Hewlett Packard and France Telecom have demonstrated their confidence in the city's future by taking stakes in local companies, support services and infrastructure.

One advantage of a multi-equilibria approach to the urban economy is that it keeps in play both stability and innovation through adaptation. A degree of continuity may be important to protect core assets, institutions and livelihoods while new opportunities are developed and the transition to new forms of production occurs. Stability gives organizations and individuals time to adapt and develop new functions, systems and skill-sets, although unwarranted protection may of course inhibit entrepreneurship and adaptation by relieving the pressure to change. There are difficult balances involved, which will vary depending on the context and the collective choices that are made.

Some evolutionary economists [23] are skeptical of a multi-equilibria framework for analyzing the economy. They acknowledge that ecological systems may attain stable states if left undisturbed, but economic systems are considered different. Economic evolution depends on the actions of individual agents who can experiment, learn and change their behavior. While economies may exhibit stability and self-organization, they are essentially about ongoing adjustment to new and emerging conditions, especially in the face of intensified international competition. This shifts the focus from a desire to stabilize conditions to an imperative to support continuous adaptation through rising productivity and innovation over time.

### 2.3.3 SOCIO-ECOLOGICAL RESILIENCE

This notion of resilience focuses on the dynamic interaction between social and ecological change [23]. The human and biophysical systems are seen as linked and co-evolving rather than independent. This differs from the previous perspectives that focus on recovery to a state where economic, social and environmental relationships are stable, or shift to a new stable domain. Communities and cities are complex human systems that interact with a variety of natural systems operating at different levels [12]. These interactions occur across multiple scales, creating a situation in a state of great flux. A simple example is the dependence of contemporary urban populations on food produced in many different rural ecosystems, some of which are within the same region and others in different countries and continents. An incidental benefit of having diverse suppliers and back-up systems ('spare capacity') is enabling the core function of food security in the city to be maintained if one of its suppliers fails.

The key features of complex adaptive systems are described by Cilliers [26] as having many elements that interact dynamically with each other, creating direct and indirect feedback loops. The behavior of the system is better explained by the nature of the interactions than by a focus on the components of the system alone. The concept of 'emergence' is used to describe how surprising patterns arise out of many relatively simple interactions and their ripple effects. Emergence does not disregard causality, but it argues against linear relationships and deterministic prediction. Feedback loops and reinforcing mechanisms make it difficult to forecast the evolving behavior of complex systems, even with full information [20]. Such systems also have self-organizing capabilities, which give them strength and the ability to transform in the face of internal and external threats.

The socio-ecological notion of resilience treats it as a process rather than the description of an outcome [7]. The focus is not a state of equilibrium but a state of continual adjustment and evolution. This gives rise to the metaphor of the adaptive cycle. This has four distinct phases in the structure and function of a system: growth or exploitation (r), conservation or consolidation (K), release or creative destruction ($\Omega$) and renewal or reorganization ($\alpha$) [27]. In the first phase, resources and assets are developed and the system stabilizes. Then a slower conservation phase occurs where the system becomes more predictable and brittle. This opens up new and uncertain possibilities, implying that as systems mature they become less resilient and more fragile. Consequently, this is followed by systemic breakdown when resources are released, and then a new phase of reorganization and regeneration occurs. This implies that crises are times of innovation and transformation, when problems can be turned into opportunities, with foresight and preparation. It is not a fixed cycle since the system can move through different sequences.

The rise and fall of adaptive cycles in the public and private sectors is well illustrated in the city of Detroit. Major car manufacturers seem to have experienced something of a comeback after a difficult period, and are re-investing in the city [28]. Their resurgence has been matched by a range of out of town investors and young entrepreneurs taking advantage of low property prices within the city. Meanwhile, a financial crisis in local government has forced the municipality to look to the federal state for assistance. In essence, the slump appears to have opened up opportunities for newcomers, social activists and entrepreneurs.

Holling [29] identifies three central properties of the adaptive cycle that shape the responses of people, ecosystems and agencies to threats and crises: the inherent potential of the system, the internal connectedness between the variables, and the adaptive capacity or resilience of the system. As the phases of the adaptive cycle progress, resilience contracts and then increases. In the disintegration phase the system is not strongly connected, so it is considered reasonably resilient. Creativity and experimentation occur because the costs of system failure are low. The variability of resilience within the adaptive cycle allows for periods of relative stability, flexibility and creative dynamism.

Recent research on London and other leading European cities has demonstrated the economic value and strength derived from dense local networks of business suppliers and services [30,31]. In a context of fast-changing markets and technologies, there is a premium on flexibility, especially as leading companies tend to be leaner and more reliant on buying-in goods and services rather than in-house production. The diversity of big cities enables firms to 'mix and match' their inputs and alter their workforce more easily in response to shifting business needs. This self-organizing, dynamic property of these places lowers costs, raises productivity and improves adaptation. Knowledge-intensive firms also benefit from superior flows of ideas and information, resulting in more learning and innovation. This enables high cost cities to differentiate themselves from competitors by continually developing more valuable products, processes and services [24]. Firms can compare, compete and cooperate, engendering a self-reinforcing dynamic that spurs progress, attracts mobile capital and talent, and generates growth from within. The most successful cities operate as knowledge hubs or gateways in a more interconnected global system of information, trade and financial flows [32].

Panarchy describes the multi-scalar nature of socio-ecological systems as they develop, disintegrate and re-emerge in response to changing conditions in different levels. The interactions between levels occur in both directions. The strength of these influences varies at different phases of the adaptive cycle. This general framework allows one to conceptualize how disruptions to systems can occur from external forces as well as internal sub-system pressures. Resilience depends on the complex interplay of these different systems, each of which under-go their own dynamic adjustment processes. This perspective helps to avoid examining individual urban areas in isolation. Cities need to be understood as dynamic systems comprised of smaller sub-systems and forming part of larger national and international systems, all of which affect their resilience capabilities [33,34].

A regional comparative study of long-term economic change in Cambridge and Swansea in the UK demonstrated how different choices and partnerships made within each city affected their levels of resilience through two recessions [20]. To boost their economy, Swansea focused on attracting foreign direct investment (manufacturing Japanese electronic products), and thereby relied on exogenous knowledge. Cambridge, on other hand, chose to invest in endogenous knowledge and build a science park, combining this with market-driven entrepreneurship and commercial exploitation of university intellectual property rights. Swansea emerged far less successfully, because their technologies became outdated and the multi-national companies did not develop local capabilities to anything like the same extent.

Swanstrom [23] draws parallels between cities and ecosystems in describing how they have extensive feedback loops that operate on multiple dimensions of time and place. In order to respond effectively to the pervasive pressures of traffic congestion, immigration and environmental degradation, cities need to be viewed as complex adaptive systems. This helps to caution against narrow, insular policy interventions that may produce unintended consequences for other parts of the system. For example, cities that seek to alleviate poverty by subsiding the travel costs of poor communities living on the periphery may inadvertently perpetuate fragmented spatial development patterns and prolong inefficient transport and bulk infrastructure arrangements, rather than encourage a more compact and integrated urban form [35].

Two additional terms can be introduced to explain the types of change that are considered essential for social-ecological resilience: adaptability and transformability [8,29]. Adaptability focuses on the ability of the socio-ecological system to adjust to external and internal change through self-organization and collective learning. Transformability is the capacity to progress to a new arrangement when the current situation is untenable and unsustainable. Socio-ecological resilience tends to place greater emphasis on the latter, i.e. systemic change to avoid getting locked-in to inappropriate structures.

## 2.4 SHIFTING THE EMPHASIS IN RESILIENCE THINKING

All three interpretations of resilience offer valuable insights for the development of sustainable cities. Their contributions recognize different features and time-scales, respond to different kinds of shocks, and exhibit varying levels of moral deliberation. Engineering resilience draws attention to the importance of short-term stability in essential urban infrastructure when extreme events like earthquakes or hurricanes occur. The moral deliberation that occurs tends to focus on those most at risk, who should receive priority attention. Because of its emphasis on efficiency in returning to a previous equilibrium and retaining the basic structure and ways of functioning, bounce-back resilience tends to leave little room for reflection on whether this position is still appropriate and desirable.

Multi-equilibria resilience has a longer-term perspective, which allows for more consideration of alternative futures. It identifies critical thresholds within the system, which warn about approaching limits to stability. The notion of multiple equilibria encourages consideration of different outcomes so as to improve conditions for current communities. Yet the focus on stable states plays down the significance of ongoing adjustment and flexibility to accommodate unforeseen pressures and crises.

Socio-ecological resilience represents a broader interpretation of resilience for more complicated, interdependent systems open to all kinds of stresses and severe events. It represents the most extended form of moral deliberation because it attempts to develop capacity within the economy, society and ecology to cope with many different factors and forces. Systems need to allow for continual adaptation to changing circumstances and inherent instability. Coping with the uncertainty, unpredictability and risks associated with such a state of flux is particularly difficult for political decision-makers.

While these notions of resilience offer a range of options to deal with change (recovery, adaptation and transformation), a sharper focus on the human capabilities as well as the technological aspects of change would further strengthen its analytical power. This can be achieved by drawing on the insights of vulnerability theory and transition theory more strongly. The following section provides further insights from this perspective.

## 2.5 DEVELOPING THE HUMAN FACE OF A RESILIENT CITY

Resilience and vulnerability are two different, yet related, ways of framing responses to social-ecological change [36]. Exactly how they relate is contested. Some view vulnerability as the opposite of resilience, while others see vulnerability as a component of resilience. It is an oversimplification to treat resilience as the converse of vulnerability because they refer to slightly different features [37]. Resilience is the responsiveness of the system, i.e. its elasticity or capacity to rebound after a shock, indicated by the degree of flexibility, persistence of key functions, or ability to transform. Vulnerability is more about the susceptibility of the system or any of its constituents to harmful external pressures.

The scope of both frameworks also tends to differ, partly because they have different origins and research traditions. Resilience emerged from ecology, whereas vulnerability emerged from political ecology, political economy and disaster risk approaches [36]. Traditionally resilience has tended to emphasize the ultimate impact on biophysical ecosystems, whereas vulnerability was oriented towards human systems and social outcomes [36,38,39]. Vulnerability theory has traditionally paid more attention to the values and agency of stakeholders along with issues of socio-historical change, identifying who is responsible for supporting marginalized people and places so they are not left behind. Resilience has focused on system dynamics and interconnections, ecological thresholds and feedback loops. In the past, it addressed the human dimension largely in the context of managing resources and ecosystem services [36]. However, as resilience theory has become used more frequently in the social sciences, questions about the resilience of what and for whom have become more central to the debate [40]. These issues determine what desirable functions of ecosystems are given priority in what areas of need. Previously the domain of vulnerability theory, the plight of marginal and disenfranchised groups is now part of resilience analysis.

In the past, resilience thinking has been criticized for not questioning the social and economic systems that cause inequality and marginalize communities enough [23,41,42]. This criticism is perhaps still justified in the case of engineering resilience because of its emphasis on preserving the status quo. It is not so valid, however, with other interpretations of resilience, which offer greater scope to consider alternative social realities and outcomes. Resilience theory may also have been conservative in concentrating on how existing social systems can absorb change and retain their essential structure and functions, rather than how they can be reformed or transformed [12,43].

The body of literature on vulnerability has engaged with these kinds of issues more extensively than the resilience literature [37]. This may be attributable to the more diverse disciplines that have influenced vulnerability thinking, including political ecology and political economy. Some concern has been expressed that the resilience framework applied to social systems neglects the man-made character of the laws and other institutions governing society. Established rules and frameworks tend to be taken for granted without questioning their rationale and validity [44]. An unintended effect may be to sanction the persistence of social injustice or environmental harm, or at least to play down the possibility of substantive change.

These kinds of concerns are starting to be raised in the resilience literature. Moore and Westley [45] discuss the importance of networks and social innovation for resilient systems. They argue that institutional entrepreneurs with specific skill-sets can be key agents in bringing about change. By establishing the right kinds of relationships that expand their networks at the right time, they can bridge divides and affect change at a larger scale.

In summary, insights from vulnerability thinking can complement resilience perspectives by drawing greater attention to human systems, social outcomes and the role of political decision-making. They remind researchers to avoid an uncritical acceptance of the underlying assumptions about the society and its structures of political and economic power. The next section considers the insights from transition theory.

## 2.6 INTRODUCING TECHNOLOGICAL CHANGE

The primary focus of resilience on socio-ecological systems has tended to relegate technology to an external factor [46]. There is a case for incorporating technological change more directly into resilience thinking because of its role in enabling change and building resilience capabilities. While social processes shape the development of technology and its absorption by households and firms, technologies also create the possibility for new social and environmental practices. Indeed technology of one sort or another is vital to the interactions within almost all socio-ecological systems and exerts a major influence on their outcomes. Technology can help to re-engineer cities so as to reduce their ecological burden and impact on natural resource consumption by making more efficient use of scarce resources, enhancing local capacity to produce food and energy, and reducing the amount of waste they generate through recycling.

This lies behind the growing interest in the notion of 'green urbanism', or a new paradigm for building more sustainable cities [47,48]. The vision goes beyond piecemeal environmental initiatives to suggest a perspective of cities as more self-contained systems in which a higher proportion of food, building materials, energy and other resources consumed are procured locally. One element would include clean production technologies and transport systems, which use fossil fuel energy resources more efficiently and generate lower carbon emissions than older methods. Another would be new techniques of energy conservation and green building to save running costs for firms and households, and create jobs from construction and retrofitting. A third ingredient would be improved systems of environmental protection and waste management to restore degraded ecosystems and support enhanced livelihoods for poor communities. Closing resource loops means shifting from a linear to a circular urban metabolism, where waste flows are redefined as productive inputs to other urban activities. A green agenda of this kind could create all sorts of opportunities to establish stronger local economies, healthier lifestyles and more livable places with smaller ecological footprints.

Transition theory uses a multi-level framework to analyze the interaction between technological and social change [49,50]. This helps to explain how technical innovation influences socio-ecological processes. It analyses transitions through three hierarchical structures: niches, regimes and landscapes [51]. A regime is the institutional and infrastructural arrangement within which a particular technological system functions. It emerges through the everyday interactions between actors and institutions and is embodied in a wide-ranging set of engineering practices, production processes, institutional procedures and skill-sets. Changes at regime level tend to be incremental without radically altering traditional practices and production systems.

The landscape is the level above this, encapsulating the broader political, social and cultural structures that form part of the bedrock of society. They are highly resistant to change and may block developments that threaten established interests and power relations. Niches represent the diverse spaces within regimes that have some protection from prevailing institutional practices, market forces, social norms and/or regulatory standards. These spaces facilitate social experimentation and technological innovation because some of the restrictive conditions that exist in the rest of the regime are relaxed or do not apply.

Masdar City may be an example of an attempt to foster niche innovation in a landscape of diminishing fossil fuels and global warming. It is being planned by the oil-rich state of United Arab Emirates (UAE) as an "eco-city" experiment in Abu Dhabi [52]. Masdar City is designed to accommodate 40,000 residents and 50,000 daily commuters in carbon neutral, zero waste conditions. The project is intended to pilot and develop the full range of specialized skills, innovative building techniques, renewable energy systems and infrastructure networks to realize the eco-city concept. An Institute of Science and Technology has been established in collaboration with MIT in the US to facilitate the two-way transfer of knowledge and expertise between the niche of Masdar City, the wider context of the UAE and other parts of the world.

Transition theory adds a further dimension to resilience in that it transcends the physical setting of a socio-ecological system. It recognizes that technological change may occur across multiple locations through organized and spontaneous flows of information and expanding communication networks [46]. In other words, the diffusion of technology across and between systems can be an important part of transformation. Access to creativity and innovation occurring in other cities and nations extends the options available for local regeneration and restructuring. This depends on the relevance of the technology designed elsewhere to the local context and the existence of sufficient institutional capacity and resources to absorb the lessons learnt [53].

An important aspect of keeping an urban economy vibrant is being receptive to new technological innovation elsewhere and finding ways of attracting and also developing these skills within the city [54]. One of the ways of achieving this is by developing innovation systems within local institutional structures that are focused both on the vulnerabilities within an urban area as well as the possible solutions for these problems found elsewhere. Attempts to achieve this in the past have included the establishment of science parks, technology incubators and living laboratories.

Summing up, transition theory extends resilience thinking by incorporating technological change more directly into the analysis of socio-ecological systems. This also implies paying greater attention to the connectivity between cities and other systems, including the flows of information, knowledge and finance. These flows can perform a positive and progressive function, or they can expose cities to greater risks and instability.

## 2.7 EXPANDING THE ANALYTICAL FRAMEWORK FOR RESILIENCE RESEARCH

Recent reports on developing resilience in cities have largely focused on disaster and risk management [14,15]. Their analytical framework, though open to concept of adaptation, is mostly limited to shorter-term resilience in the face of natural disasters like flooding and drought. They analyze, for example, the specific resilience of the existing infrastructure in a city to natural disasters, as opposed to the general resilience of its infrastructure towards longer term issues of transformation. The importance of specific or bounce-back resilience is undisputed as there is a need to ensure that vital facilities like power plants, water and sanitation systems and hospitals resume functioning as soon as possible following a disaster. Without this kind of first-order resilience, a society and its supporting economy can readily be brought to its knees.

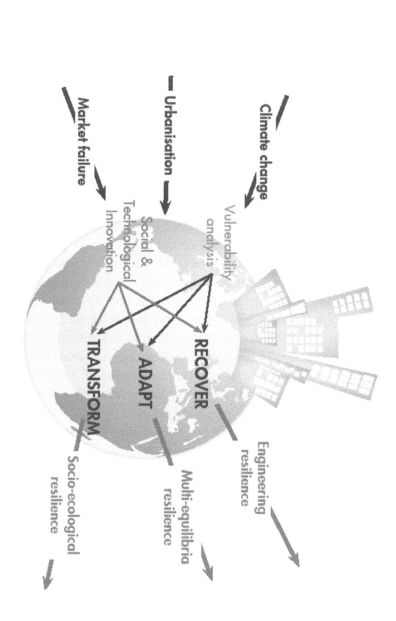

**FIGURE 1:** Dimensions of a resilient city.

This paper attempts to broaden the concept of resilience by expanding the analytic framework for understanding sustainable cities. In addition to examining engineering resilience it discusses two other kinds of resilience related to the long-term sustainability of cities as dynamic social organizations: multi-equilibria resilience and socio-ecological resilience. We also introduce two other aspects of resilience: human vulnerability and innovation which though discussed within the literature, are not emphasized as much as they are in the vulnerability and transition theory analyses. The diagram (Figure 1) below serves as an illustration of the discussion that follows on the different the types of resilience and their relationship to vulnerability studies and social or technological innovation.

Multi-equilibria resilience emphasizes the importance of paying attention to tipping points or thresholds within social, economic and ecological systems to avert disaster. This second dimension of resilience could be referred to as preventative and adaptive resilience where warning signs in social, economic and ecological environments are monitored so that interventions can be planned before crises arise and the entire socio-economic and ecological system is weakened or seriously damaged. Research focusing on this type is already found in attempts to measure resilience in city economies by producing ecological and economic indicators to benchmark levels of water purity, air pollution, unemployment, crime and education. These indicators are used to monitor critical levels to enable adaptation or intervention before serious damage occurs.

Socio-ecological resilience is a third dimension that looks beyond such indicators at the dynamic interchange between the social and ecological factors in an economy and their simultaneous interaction with these same forces both beyond the city borders and within various precincts within the city. It speaks about a more general form of resilience that is aimed at long term stability and flexibility. In this third dimension, what are being gauged are essentially the deep underlying forces driving change both within the economic system and outside it. The focus is on understanding the kind of long-term forces at play rather than attempting to predict particular outcomes in the short term. The value of a more general resilience analysis lies in understanding long-term adaptive capacity to cope with constant change.

This paper calls for two additional areas of background analysis in resilience theory, one aimed at identifying and supporting processes of social innovation in vulnerable communities, and the other at identifying processes of technological innovation. In order to achieve a better understanding of the former, resilience analyses are likely to benefit greatly from incorporating the insights of vulnerability studies. Understanding the particular challenges and capabilities of the people most affected by change is likely to improve their ability to adapt or transform it. If resilience analyses incorporate vulnerability insights, they would be better equipped to identify processes of social progress within these communities and support the social entrepreneurs who are trying to bring about institutional change. Secondly, identifying and supporting processes of technological innovation would also extend resilience. While resilience theory makes explicit reference to transformation within socio-ecological resilience, the focus remains socio-ecological cycles rather than socio-technical cycles. Socio-ecological interactions are physically located, whereas socio-technical systems have both local and global bearings. If resilience theory were to incorporate the insights of transition theory in their analyses of change within socio-ecological systems, they would strengthen their analytical capabilities. The core focus at this level of resilience analysis is an understanding of how an urban area nurtures and attracts technological innovation. It demands an assessment of relationships between government, economic and social agents with many practical implications.

It is proposed that these three dimensions of resilience and two background analyses are critical for identifying the core features of building sustainable cities. These core features are summarized as: the ability to recover, to adapt to new equilibria, to transform when necessary and to facilitate social and technologically innovative processes. How this can be achieved within a particular city and its unique social and ecological adaptive cycles would need to be researched within each particular context.

## 2.8 CONCLUSIONS

The paper has reviewed several different theoretical frameworks for understanding the essential features of sustainable cities—resilience, vulnerability and transitions. The broad perspective of resilience provides a range of insights into how major urban areas deal with external challenges and shocks to the way they manage their assets and resources. The fundamental point is that for the simplest objectives of survival and growth, cities need to be able to avoid hazardous conditions where possible or respond positively where risks are unavoidable and change is inevitable.

In some circumstances they need above-all to withstand disasters and bounce back from adverse stress (engineering resilience). Retaining essential urban functions is important to avoid systemic crises and breakdown. In other situations it is vital that cities adapt to a new environment by adjusting certain structures and systems to match the altered conditions (multi-equilibria resilience). They may be experiencing stagnation or have reached a critical threshold beyond which incremental change is necessary to regain a steady state. In a third, highly dynamic situation, a process of ongoing change occurs as cities continually rearrange their structures and reorganize their institutions (socio-ecological resilience). It is important for them to avoid becoming locked into a single development path based on outdated arrangements and inappropriate patterns of resource use.

Resilience thinking initially tended to neglect how socio-ecological change affected a particular community. Engineering approaches to resilience especially tend to play down the diverse interests within cities and the need to disentangle the ultimate objectives of sustainability and resilience—for whom and for what? Theories of vulnerability can complement resilience by elevating the importance of human systems and social outcomes. They caution against an uncritical acceptance of the underlying power relations, and encourage more explicit consideration of the role of political decision-making. Transition theories focus on the relationship between social and technological change. They can help resilience research to guard against inertia and point to the importance of knowledge and information flows between different cities, nationally and internationally.

Previous research has tended to portray these theoretical frameworks as alternatives. Researchers and writers have typically advocated one perspective over another, and have generally applied them in isolation. The present paper has suggested that their insights could be complementary and mutually supportive. For resilience analyses to identify and support process of social innovation within communities, they are likely to need to understand the vulnerabilities and social dynamics that exist in these places. Vulnerability studies can provide this background. Similarly, if resilience analyses are to effectively identify and support processes of technological innovation in a city, an understanding of how these processes function is vital.

This paper advocates a resilience approach to building sustainable cities. This is because the multiple interpretations of resilience provide diverse insights into the broad attributes required to prepare a city for short and long term change. The paper suggests that resilience theory could improve its analyses of social and technological innovation by adopting some of the insights of vulnerability and transition theory.

The issue of when to apply a particular kind of resilience in a city depends entirely on the natural and human resources available in that place. In some circumstances the key attribute of a resilient city is its ability to recover quickly from a shock. In other circumstances cities need to adjust to new conditions because their established structures or systems are no longer fit for purpose. This process of change may be ongoing if the environment is continually shifting and it is not feasible or viable for cities to stand still.

Each type of resilience may have a somewhat different time-frame. Coping may be the most appropriate short-term response to a disaster, while adaptation is likely to be important in the longer-term. Each may also apply more to one feature of the city than to another—its physical infrastructure, public services, leading economic sectors, or governing institutions. Some aspects may require one-off or incremental adaptation, while others require continuing adjustment or enhancement. These differences may also vary for cities at different stages of development or located in different territories, so that some require all-round improvement while others need modest modification. Determining which perspective is most appropriate in what circumstances is neither simple nor straightforward.

It is the task of resilience researchers to balance and prioritize the different attributes discussed in this paper when attempting to map out resilience strategies for cities. While there may be tensions and contradictions between the different perspectives that cannot be neatly reconciled in practice, cities that are capable of maintaining stable, yet flexible and innovative systems stand the best chances of developing sustainably.

## REFERENCES

1. World Urbanisation Prospects; United Nations: New York, NY, USA, 2012.
2. Grant, R.; Thompson, D. The development complex, rural economy and urban-spatial and economic development in Juba, South Sudan. Local Econ. 2013, 28, 77–88.
3. Holling, C.S. Resilience and stability of ecological systems. Annu. Rev. Ecol. Syst. 1973, 4, 1–23.
4. Brand, F.S.; Jax, K. Focusing the meaning(s) of resilience: resilience as a descriptive concept and a boundary object. Ecol. Soc. 2007, 12, 23:1–23:16.
5. Carpenter, S.; Walker, B.; Anderies, J.M.; Abel, N. from metaphor to measurement: Resilience of what to what? Ecosystems 2001, 4, 765–781.
6. Christopherson, S.; Michie, J.; Tyler, P. Regional resilience: Theoretical and empirical perspectives. Camb. J. Reg. Econ. Soc. 2010, 3, 3–10.
7. Pendall, R.; Foster, K.A.; Cowell, M. Resilience and regions: Building understanding of the metaphor. Camb. J. Reg. Econ. Soc. 2010, 3, 71–84.
8. Folke, C.; Carpenter, S.R.; Walker, B.; Scheffer, M.; Chapin, T.; Rockström, J. Resilience thinking: Integrating resilience, adaptability and transformability. Ecol. Soc. 2010, 15, 1–9.
9. Carpenter, S.R.; Arrow, K.J.; Barrett, S.; Biggs, R.; Brock, W.A.; Crépin, A.S.; Engström, G.; Folke, C.; Hughes, T.P.; Kautsky, N.; et al. General resilience to cope with extreme events. Sustainability 2012, 4, 3248–3259.
10. Evans, J.P. Resilience, ecology and adaptation in the experimental city. T.I. Brit. Geogr. 2011, 36, 223–237.
11. Davoudi, S. Resilience: A bridging concept or a dead end? Plann. Theor. Pract. 2012, 13, 299–307.
12. Cote, M.; Nightingale, A.J. Resilience thinking meets social theory: Situating social change in socio-ecological systems (ses) research. Prog. Hum. Geogr. 2012, 36, 475–489.
13. Pelling, M. Adaptation to Climate Change: From Resilience to Transformation; Routledge: London, UK, 2011.
14. United Nations Office for Disaster Risk Reduction (UNISDR). Making Cities Resilient Report; UNISDR: New York, NY, USA, 2012.
15. Godrey, N.; Savage, R. Future Proofing Cities: Risks and Opportunities for Inclusive Urban Growth in Developing Countries; Atkins: Epsom, UK, 2012; p. 188.
16. Folke, C. Resilience: The emergence of a perspective for social-ecological systems analyses. Global Environ. Change 2006, 16, 253–267.
17. United Nations Disaster Risk Making Cities Resilient Campaign. Available online: http://www.unisdr.org/campaign/resilientcities/cities/view/2764/ (accessed on 10 April 2013).
18. Adger, W.N. Social and ecological resilience: Are they related? Prog. Hum. Geogr. 2000, 24, 347–364.
19. Noya, A.; Clarence, E. Community capacity building: Fostering economic and social resilience. Organisation for economic cooperation and development, 26–27 November 2009. Available online: http://www.oecd.org/dataoecd/54/10/44681969.pdf?contentId=44681970 (accessed on 10 April 2013).

20. Simmie, J.; Martin, R. The economic resilience of regions: Towards an evolutionary approach. Camb. J. Reg. Econ. Soc. 2010, 3, 27–43.
21. Adams, D.; Tiesdell, S. Shipping Places: Urban Planning, Design and Development; Routledge: London, UK, 2012.
22. Robinson, B.M. Optimizing infrastructure: Urban patterns for a green economy; UN Habitat: Nairobi, Kenya, 2012.
23. Swanstrom, T. Regional Resilience: A Critical Examination of the Ecological Framework; Working Paper 2008-7; Institute of Urban Regional Development, University of California: Oakland, CA, USA, 2008.
24. Turok, I. The distinctive city: Pitfalls in the pursuit of differential advantage. Environ. Plann. A 2009, 41, 13–30.
25. The Economist. Upwardly mobile: Kenya's technology start-up scene is about to take off, 2012. Available online: http://www.economist.com/node/21560912 (accessed on 18 November 2012).
26. Cilliers, P. What can we learn from a theory of complexity? Emergence 2000, 2, 23–33.
27. Walker, B.; Holling, C.S.; Carpenter, S.R.; Kinzig, A. Resilience, adaptability and transformability in social-ecological systems. Ecol. Soc. 2004, 9, 5.
28. Davey, M. A private boom amid Detroit's public blight. New York Times, 4 March 2013. Available online: http://www.nytimes.com/2013/03/05/us/a-private-boom-amid-detroits-public-blight.html?pagewanted=all/ (accessed on 10 April 2013).
29. Holling, C.S. Understanding the complexity of economic, ecological, and social systems. Ecosystems 2001, 4, 390–405.
30. Buck, N.; Gordon, I.; Harding, A.; Turok, I. Changing Cities: Rethinking Urban Competitiveness, Cohesion and Governance; Palgrave: London, UK, 2005.
31. Hall, P.; Pain, K. The Polycentric Metropolis; Earthscan: London, UK, 2006.
32. Castells, M. The Information Age: Economy, Society and Culture. Volume 1: The Rise of the Network Society, 2nd ed.; Blackwell: Oxford, UK, 2000.
33. Lucci, P.; Hildreth, P. City Links: Integration and Isolation; Centre for Cities: London, UK, 2008.
34. Turok, I. Cities, Regions and competitiveness. Reg. Stud. 2004, 38, 1069–1083.
35. Behrens, R.; Wilkinson, P. Housing and urban passenger transport policy and planning in South African cities: A problematic relationship? In Confronting Fragmentation: Housing and Urban Development in a Democratising Society; Harrison, P., Huchzermeyer, M., Mayekiso, M., Eds.; Cape Town University Press: Cape Town, South Africa, 2003; pp. 154–174.
36. Miller, F.H.; Osbahr, E.; Boyd, F.; Thomalla, S.; Bharwani, G.; Ziervogel, B.; Walker, J.; Birkmann, S.; van der Leeuw, S.; Rockström, J. Resilience and vulnerability: Complementary or conflicting concepts. Ecol. Soc. 2010, 15, 11.
37. Turner, B.L. Vulnerability and resilience: Coalescing or paralleling approaches for sustainability science? Global Environ. Change 2007, 20, 570–576.
38. Eakin, H.; Luers, A.L. Assessing the vulnerability of social-environmental systems. Annu. Rev. Environ. Resour. 2006, 31, 365–394.
39. Ernstson, H.; van der Leeuw, S.E.; Redman, C.L.; Meffert, D.J.; Davis, G.; Alfsen, C.; Elmqvist, T. Urban transitions: On urban resilience and human-dominated ecosystems. AMBIO 2010, 39, 531–545.

40. Porter, L.; Davoudi, S. A politics of resilience for planning: A cautionary note. Plann. Theor. Pract. 2012, 13, 329–333.
41. MacKinnon, D.; Derickson, K.D. From resilience to resourcefulness: A critique of resilience policy and activism. Prog. Hum. Geog. 2012, 37, 253–270.
42. Hassink, R. Regional resilience: A Promising concept to explain differences in regional economic adaptability? Camb. J. Reg. Econ. Soc. 2010, 3, 45–58.
43. Hudson, R. Resilient regions in an uncertain world: Wishful thinking or a practical reality? Camb. J. Reg. Econ. Soc. 2010, 3, 11–25.
44. Christmann, G.; Ibert, O.; Kilper, H.; Moss, T. Vulnerability and Resilience from a Socio-spatial Perspective. Towards a Theoretical Framework; Working Paper for Leibniz Institute for Regional Development and Structural Planning (IRS); No. 45; IRS: Leibniz, Germany, 2012; pp. 1–31.
45. Moore, M.; Westley, F. Surmountable chasms: Networks and social innovation for resilient systems. Ecol. Soc. 2011, 16, 5.
46. Smith, A.; Stirling, A. The politics of social-ecological resilience and sustainable socio-technical transitions. Ecol. Soc. 2010, 15, 11.
47. Beatley, T., Ed.; Green Cities of Europe; Island Press: Washington, DC, USA, 2012.
48. Swilling, M.; Annecke, E. Just Transitions: Explorations of Sustainability in an Unfair World; UCT Press: Cape Town, South Africa, 2012.
49. Haxeltine, A.; Seyfang, G. Transitions for the People: Theory and Practice of 'Transition' and 'Resilience' in the UK's Transition Movement; Tyndall Working Paper 134; University of East Anglia: Norwich, UK, 2009.
50. Van der Brugge, R.; van Raak, R. Facing the adaptive management challenge: Insights from transition management. Ecol. Soc. 2007, 12, 33.
51. Foxon, T.J.; Reed, M.S.; Stringer, L.C. Governing long-term social-ecological change: What can the adaptive management and transition management approaches learn from each other? Environ. Pol. Govern. 2009, 19, 3–20.
52. Peter, C.; Swilling, M. Sustainable, Resource Efficient Cities—Making it Happen; DTI/1538/PA; UNEP Division of Technology Industry and Economics (DTIE): Paris, France, 2012.
53. Coyle, D. The Economics of Enough: How to Run the Economy as if the Future Matters; Princeton University Press: Princeton, NJ, USA, 2011.
54. Ridge, S.G.M. Innovation, innovation systems, science parks and living labs. Unpublished paper for the South African Department of Science and Technology. 2010.

# CHAPTER 3

# Resilience Attributes of Social-Ecological Systems: Framing Metrics for Management

DAVID A. KERNER AND J. SCOTT THOMAS

## 3.1 INTRODUCTION

What makes a system resilient, and how can we manage for this condition?

In order to build and maintain resilience, managers must be able to understand what qualities or attributes enhance—or detract from—a system's resilience. We present, herein, a set of resilience attributes that enable policy makers, program managers, and stakeholders in collaborative initiatives to make practical use of resilience theory in understanding and managing their systems of interest.

We offer these resilience attributes as a basis for exploring and enhancing a system's resilience; they can be employed by leaders within a system, collaboratively by stakeholder groups, and by those having responsibility for subsystems within the whole. It is intended to enable stakeholders at all levels of a system, including the subsystem and component levels, to readily assess and improve the resilience of their part of the whole. It is also intended to promote, through application, a basic understanding and enduring awareness of resilience concepts and their role in system function.

These standardized resilience terms, from which metrics can be derived, can be incorporated into resource management plans and decision-support tools to help managers assess the current resilience status of their systems, gauge progress toward goals, make rational resource allocation decisions, justify funding choices, and enable training and unity of purpose across programs. The resilience attributes can also facilitate deeper investigations of resilience through ongoing efforts to model system behavior, and more importantly, they provide an approachable set of terms for non-specialists to incorporate into plans and programs lacking access to (or the desire to use) sophisticated models. Our intention is to provide policy makers and resource managers practical access to resilience theory for work in the real world.

It is important to clarify terminology up front. Much research has been performed in recent decades regarding resilience theory, and there are competing definitions for "resilience." One definition of resilience is the amount of disturbance a system can resist or the speed with which it returns to equilibrium. This definition of "engineering resilience" [1] is useful for describing closed systems, but many systems are not closed, limiting the usefulness of this approach.

Business leaders may think of resilience as "the capacity of an enterprise to survive, adapt, and grow in the face of turbulent change" [2]. However, current best practice in business and government usually consists of optimizing the production and delivery of goods and services [3]. Pursuing efficiency often requires tight control of a system's elements in isolation to create a steady "maximum sustainable yield." This approach is not necessarily sustainable in unpredictable, "open" systems that must adapt to a wide range of external perturbations—that is, most systems we find ourselves working within and attempting to manage.

Natural resource managers observe that ecosystems adapt, survive, and thrive despite a wide range of stresses and disruptions. Systems that are characterized by uncertainty and unpredictability appear more tractable when examined from an ecological systems perspective. It should be noted, though, that "natural" systems do not operate without human influence (even the most remote ecosystems feel the effects of climate change, industrial pollution, and aerosol-induced ozone depletion), and "human" systems do not function unaffected by the natural environment. This complex interconnectedness of humans and their environment is embodied in what are called social-ecological systems, or SESs [3]. For the purposes of this paper, an SES may be of any size, complexity, origin, or purpose, each with its own unique, highly varied linkages. An SES may consist of a community and the commercial, industrial, agricultural, water and energy supply, and other ecosystem services upon which it depends. Scaling down within that community it might be the local fishing industry and the fishery upon which it depends, or a factory and the workers, managers, suppliers, consumers, and resources and waste disposal associated with that factory.

We are concerned herein with the ways humans can assess and affect the resilience of an SES. For that purpose, we have oriented our efforts to satisfy the needs of people who normally focus on the systems, not on resilience concepts. In this context, resilience has been defined as the capacity of a linked SES to experience shocks while retaining essentially the same function, structure, feedbacks, and therefore identity [1,3,4,5,6]. How much disturbance can the system—people, infrastructure, resources, and environment—accommodate while still maintaining its basic structure, capabilities, and capacity to function? How far can it bend and adapt without breaking? It is this definition we use as a basis for describing resilience attributes of systems in order to manage for system resilience.

Systems often fail in unpredictable ways, but resilient systems continue to function despite the challenges. Planning assumptions do not always hold true, and planners often fail to ask how well the system will function in the face of large, unexpected challenges, or the accumulation of many smaller stresses; these include market failures; geopolitical and demographic shifts; resource shortages (e.g., water, fuel, fertilizer, minerals); epidemics; climate change; and technology disruptions [7,8]. Retention of the SES identity is a concern since an SES can exist in alternate regimes or "stability domains," some less desirable than others. There are thresholds that, once passed, invite different system feedbacks leading to changes in function and structure [1,3,5,9]. Social-ecological systems can sometimes get trapped in very resilient but undesirable regimes. Adaptation may not be possible, and escape may require significant energy to cause change [10]. Thus, "the more resilient a system, the larger the disturbance it can absorb without shifting into an alternate regime" [5]. Being able to understand which resilience attributes of a managed SES need attention is an important first step toward avoiding undesirable thresholds, absorbing shocks, mitigating disruptions, and managing transitions.

## 3.1.1 THE NEED TO TRANSLATE THEORY INTO PRACTICE

Resilience theory provides a powerful, holistic paradigm for understanding system dynamics. And there appear to be a number of conceptual strategies for enhancing resilience within a system. Some strategies involve (1) manipulating an important system variable or "reshaping the basin of attraction or stability domain" to reduce the odds of breaching thresholds [6,11,12,13]; (2) moving the "ball" away from the walls of the basin (moving the system away from a threshold) [6,14]; (3) understanding and manipulating the focal system's position within the Adaptive Cycle [3,15,16]; (4) manipulating factors at a different scale (or in a connected system) to increase system resilience at the focal scale (panarchy) [12,14,16,17]; (5) assessing the system for strengths and vulnerabilities based on attributes of system resilience and work to eliminate weaknesses. All of these strategies hold promise, and there are no doubt other strategies yet to be proposed. However, if it is to be of practical value for policy makers and resource managers, the theory must be translated into sensible decision-support tools universally applicable across large and small systems and enterprises. Towards that end, option five appears to be a direct, less complicated, and administratively tenable method suitable for those who want to incorporate resilience management into strategic planning as well as for those who do not wish to rely upon consulting experts, modelers, and large budgets.

As noted by Fiksel, "There is a great need for operational definitions and metrics for sustainability and resilience in economic, ecological, and societal systems" [2]. The resilience attributes presented herein are in answer to that need, providing a means for applying resilience concepts in tangible applications. Discussion of the theoretical underpinnings of these resilience attributes and the methodology and results of a resilience assessment based on the system attributes will be the subject of forthcoming articles; the focus of this article is to provide useful information for the non-specialist to apply to systems or programs that they manage and that would benefit from a resilience perspective.

Senior federal agency decision-makers have called for the means to instill and manage resilience in their programs; frustrated with inadequate, incompatible, and uncoordinated management policies, strategies, and tools that cannot accommodate resilience analysis, they have openly sought resilience metrics that would enable prioritization of activities and allocation of resources [7,18,19,20]. For example, the National Defense Industrial Association publishes each year the top issues it recommends DoD and the defense industry give their greatest attention. In its 2013 edition, it cites energy security as its third of eight issues, and specifically states that:

*"Energy resilience metrics should be part of the overall performance metric, and should be considered in requirements development and acquisition processes. This shift in emphasis from assuring supplies to assuring mission preparedness will complement and reinforce the mandate that mission performance takes priority over energy consumption. This new focus will also ensure future planning addresses not just energy supplies, but actual mission performance for the widest range of circumstances."*

NDIA concludes with the following recommendation:

*"Despite the challenge to transition to an energy strategy that incorporates resilience and adaptability to evolving conditions given planned budget constraints, shifting mission priorities, and a need for flexibility in a changing global energy reality, DoD should consider making resilience a focus for energy security".* [21]

This concern for defined attributes and metrics is understandable: Managers must monitor and measure what they manage. Since data collection and analysis is expensive, it is important to define the properties of resilience so that monitoring and assessment can be appropriately targeted. And just as important, this must be an integral component of the prevailing strategic management framework—Vision, Mission, Goals, Objectives, Metrics, and Milestones—that agencies and organizations use to set their program agendas.

Translating theory into practice, however, must yield terminology suited to the community of interest, in this case the SES stakeholders. Indeed, Gunderson and Folke have stated that "one of the major unanswered themes [is] the gap between science and policy that seems to be widening in many places. In more practical or basic terms, what is the relevance of resilience scholarship to practical issues?" [16]. That scholarship must meet the needs of those working with real-world systems and aligned with their functional perspectives. "Expert solutions may maximize something, but they rarely maximize legitimacy" [11]. For stakeholder groups collaborating in initiatives to enhance resilience, it is preferable to have a user-friendly, practical method for communicating intent, translating intent into goals and objectives, and planning implementation activities rather than to rely upon a "black-box" approach. As such, we have focused on the synthesis of terms and definitions appropriate for those whose main concern is their system, not resilience concepts.

## 3.2 RESILIENCE ATTRIBUTES

Social-ecological systems appear to share attributes by which resilience can be characterized and assessed. We have investigated which system attributes relate specifically to the ecological definition of resilience, asking, "What attributes reflect whether a system will be able to continue to function and retain its identity in the face of existential challenges?" We considered attributes for all types of systems, including natural and manmade, physical and institutional, small and large, simple and complex. A number of researchers have defined various attributes of system resilience.

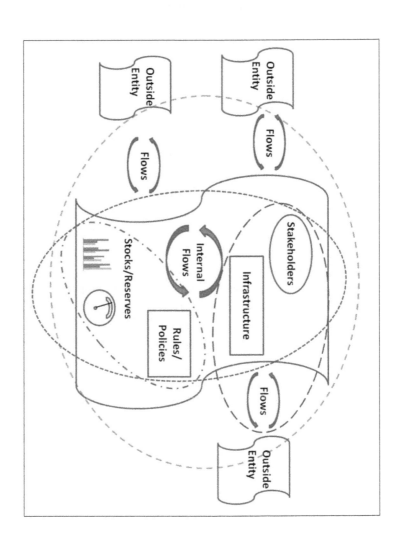

**FIGURE 1:** Extent of knowledge about SES.

### 3.2.1 PREVIOUS EFFORTS TO MOVE FROM THEORY TO APPLICATION

In his seminal work on the subject, Holling examined the concepts of resilience and stability and described how diversity and connectivity contribute to system resilience [1]. Several years later, in examining the vulnerability of the American energy system, Lovins advanced the concept of managing for resilience by proposing an approach to a design science for resilience, wherein he delineates a set of key attributes for system resilience associated with both engineered and biological systems [22]. These attributes, the descriptions of which focus on a national energy system, included: fine-grained, modular structure; early fault detection; redundancy and substitutability; optional interconnection; diversity; standardization; dispersion; hierarchical embedding; stability; simplicity; limited demands on social stability; accessibility, and reproducibility. With these attributes, Lovins made significant strides in articulating terms of resilience common to both natural and human-designed systems.

Lee examined the essential contribution of adaptive management, the capacity for institutional learning, and the requirements for leadership and collaboration. Adaptive management, in particular, was advanced by Lee, with agencies and organizations implementing adaptive management widely in various forms [23].

Becker and Ostrom examined a series of case studies to formulate general principles for organizing the types of institutions that successfully manage resources in a sustainable manner [24]. They identified eight principles: clearly defined boundaries (for both resources and resource users); proportional equivalence between benefits and costs (rules specifying resource allocation are based on local conditions); collective-choice arrangements (affected participants can influence the rules); monitoring (of resources and user behavior by accountable monitors); graduated sanctions; conflict resolution mechanisms; minimal recognition of rights to organize (outside authorities do not challenge local institutions), and nested enterprises. Becker and Ostrom differentiated the institutional rule sets required for managing renewable versus single use resources.

Walker et al. examined the relationships between three related attributes of SESs that appear to determine the future trajectory of an SES: resilience, adaptability, and transformability [14]. Regarding resilience, they described four key aspects: latitude—the maximum amount a system can be changed before it breaches a threshold and loses its ability to recover its identity; resistance—the ease or difficulty of changing the system; precariousness—how close the current state of the system is to the threshold; and panarchy—the resilience of the system at one scale is dependent on system dynamics at other scales "above" and "below." Walker et al. also discussed the concepts of Stability and Balance in regards to system resilience.

Revisiting that line of research, Folke et al. examined how adaptability captures the capacity of a SES to combine experience and knowledge to learn [25]. Folke et al. described how "resilience is often associated with diversity" [4]. Discussing the importance of adaptive management (and co-management), they maintained that management actions that build resilience are flexible and open to learning and collaboration.

Folke et al. examined ecosystems undergoing change and described the importance of maintaining functional-group diversity and functional-response diversity for the conservation of biodiversity and ecosystems [10].

Walker et al. asked "When does it make sense to build resilience and what is the best way to do it for a particular SES?" In answer, they proposed a "resilience analysis and management approach" [11]. They discuss the importance of collaboration among stakeholders in the system to ensure the legitimacy of the effort.

Walker et al. advanced a series of concepts about resilience, several of which were proposed as attributes of resilient systems [5]: adaptability (including functional diversity, response diversity, redundancy, social capacity including leadership, social networks, trust, innovation and skills); linkages; institutions for self-determination; capital reserves (natural, social, financial, infrastructure), and learning, memory, and adaptive co-management.

Ostrom's Institutional Analysis and Development (IAD) Framework provided a means for assessing how institutional traits and characteristics and the leadership capacities (and limitations) of key actors at multiple levels within any system can influence system resilience [26]. She discussed the importance of learning capacity in the context of institutional structure. She also cautioned that differences between individual institutions may be so great as to confound simple comparison. This observation may argue for the development of system-specific resilience metrics based on generalized resilience attributes, as is discussed in this article. Ostrom also provided additional explanation of the role of leadership in [27], showing that when some members of a group possess entrepreneurial skills and are respected locally, self-organization to manage resources is more likely. She also addressed the role of norms for reciprocity in building trust that supports collaborative efforts.

Berkes examined the relationship of resilience and vulnerability in regards to natural hazard evaluation [28]. He identified a series of factors relevant to building resilience: learning to live with change and uncertainty; nurturing various types of diversity (ecological, social, political); developing learning capacity; promoting self-organization and autonomy; strengthening local institutions; building cross-scale linkages, and building problem-solving networks. Berkes described the concept of false subsidies, the need for tight coupling of monitoring and response, as well as how adaptive capacity is strengthened by adaptive management and institutional learning.

Examining the relationship between sustainability and resilience, Fiksel described strategies for enhancing ecological resilience including [2]: broaden knowledge sources; increase human ability to cope with change and uncertainty; introduce adaptive management practices; build networks (social and scientific).

In *Resilience Thinking*, Walker and Salt (2006) described a series of key attributes of resilience, including: Functional Diversity, Response Diversity, Modularity, Redundancy, Tightness of Feedbacks, Reserves, and Collaboration [3]. In *Resilience Practice*, Walker and Salt developed a framework for assessing general resilience that is focused on Adaptive Capacity, and specifically the attributes: Diversity, Openness, Reserves, Tightness of Feedbacks, Modularity, Leadership, Social Networks, Trust, and Levels of Capital Assets [6].

Olsson et al. examined a series of case studies across the globe, comparing the outcome of various management actions employed as an SES is transformed [29]. They found that leadership and shadow networks are key components for preparing for and effecting such transformations. Leadership functions include: spanning scales of governance; building knowledge; orchestrating networks; communicating understanding and reconceptualizing issues; reconciling problems; recognizing or creating windows of opportunity; promoting and stewarding experimentation as smaller scales; and promoting novelty by combining different networks, experiences, and memories.

Thomas and Kerner described the resilience enhancing potential of adaptive management and collaboration, as well as the negative influence of false subsidies, for energy security policies and programs [30].

Biggs et al. examined the current literature to identify seven principles for enhancing resilience of ecosystem services [12]: maintain diversity and redundancy (addressing response diversity, balance, and disparity); manage connectivity (including modularity and nestedness); manage slow variables and feedbacks; foster an understanding of SESs as complex adaptive systems; encourage learning and experimentation (addressing adaptive management); broaden participation, and promote polycentric governance systems (conveying modularity and functional redundancy). The first three of these principles address general SES properties and processes, while the remaining four focus on governance of the SES. The authors state that this is not a definitive set of principles, but one advanced to stimulate further discussion and refinement. They write: "there is a pressing need for a better understanding of how the principles can be operationalized."

In examining the related concepts and appropriate management context of robustness, resilience, and sustainability, Andries et al. discuss how incorporating the principles of managing for robustness can add rigor to managing for resilience [31]. They discuss a number of key attributes for resilient systems: adaptive capacity and response diversity; adaptive management and the capacity to learn; transformability; understanding cross-scale interactions and feedback systems, and collaboration.

Stokols et al. discussed how resilience theory relates to social-ecological systems theory [32]. They described the core principles of social ecology and how these contribute to community resilience. They presented a detailed description of the forms of capital available to and necessary for SESs to function. This includes Material Resources: economic/financial capital—financial assets for attaining productivity; natural capital—resources produced through natural processes; human-made environmental capital—physical resources designed and built; and technological capital—machinery, equipment, digital/communications devices. It also includes Human Resources: social capital—relationships among people that facilitate action; human capital—capacities of persons, including skills and information; and moral capital—investment of personal and collective resources toward justice/virtue.

## 3.2.2 STAKEHOLDER ANALYTICAL CONSTRAINTS

All resilience analysis methodologies are hampered by a limited knowledge about the system in question and the spatial, temporal, and contextual (environmental, socio-cultural, regulatory, and related) circumstances within which it functions. Figure 1 depicts different levels of knowledge a stakeholder might have about their SES.

Given the complex, adaptive, and stochastic nature of an SES, even a slight difference in system insight could yield a significantly different understanding of the system's dynamics, and therefore resilience, in the face of shocks and perturbations. These limits inherently constrain even the most comprehensive of analyses. As such, an incremental approach to resilience assessment may prove most productive, beginning with examining a system's resilience-specific attributes to develop a baseline understanding of the system's "resilience posture"; this in turn supports deeper investigations of resilience phenomena for which information beyond the normal purview of the stakeholder must be pursued. Moreover, the availability of common analytical terms supports better coordination in the further development of resilience theory.

It should also be noted that the varying degrees of stakeholder knowledge about a system, which may arise for a number of reasons, point to the importance of cooperation in order to work comprehensively within the system.

## 3.2.3 THE SYNTHESIS OF PRACTICAL RESILIENCE ATTRIBUTES

From the above examination of the literature, we can develop an understanding of key traits of resilient systems as conceived by researchers from diverse academic and practicing fields, including ecology, wildlife and fisheries management, water resource management, various branches of engineering, hazard mitigation, risk management, operations research, and institutional analysis.

As stated, our efforts are focused on making theory serviceable to the practitioner, namely the stakeholder seeking to understand and affect the resilience of a system. This requires consideration of certain factors: First, stakeholders usually know their systems best. An outsider may offer new ways of examining a system, but the stakeholder will have both the broadest and the most intimate knowledge about that system. Second, stakeholders have limited time and resources. The perceived value of additional analyses and actions will be weighed against more pressing requirements. And third, stakeholders pay the most attention to that which they believe can most affect their systems. Concern about other possible influences, internal or external, tends to fall off with a decrease in the perceived likelihood and consequence of that influence.

To accommodate these factors, we have sought to develop, derive, and compile resilience analysis terms that: are stated in the language of stakeholders; address easily-assessed system attributes without extensive knowledge of new theory; and promote the ready consideration of the broadest range of factors that could affect system resilience. Prior efforts have tended to yield theory-oriented terms that suit the resilience theoretician; we have sought to provide systems-oriented terms that favor the stakeholder involved in resilience analysis.

In addition to meeting stakeholder needs in their own vocabulary, there is another pragmatic reason for this approach: It is much harder to recognize, understand, and characterize the tremendous variety of systems from the resilience perspective than it is to recognize, understand, and characterize resilience concepts from those systems' perspective. Put another way, the world of possible systems is harder to delineate in resilience-theory terms than resilience theory is to delineate in systems terms.

Expanding on the terms employed by Lovins [22] and addressing open systems (i.e., SESs), and drawing on the authors' own research as well as that of others (as cited above and in the discussion of the individual resilience attributes below), we derived or synthesized from basic resilience theory literature the specific attributes associated with a system's resilience. We identified resilience concepts already expressed in terms stakeholders could easily use to measure or characterize their system, then crafted additional system-oriented terms to capture those concepts not yet described in stakeholder-friendly language, with the focus remaining on what a stakeholder can readily measure or characterize about their system. The resilience literature was revisited to ensure system attributes had been included that would support an assessment of each resilience concept. Finally, several iterations were made to combine like terms under single headings if it could be done without losing pertinent, independent system attributes. The terms are intentionally generic to support a wide variety of systems, with the understanding that stakeholders will employ them in a manner and language aligned with their unique systems.

To meet typical stakeholder needs, we focus on system traits that can be construed from information commonly available within the system, or easily obtained by those stakeholders. The terms form a baseline for resilience analysis, providing a snapshot of the system's resilience posture that could be retaken on a regular basis as part of an adaptive management strategy to maintain and enhance system resilience.

The attributes also form the basis for the first part of an iterative approach to assessing system resilience. Others have sought to capture all aspects of resilience theory in their attempts to postulate metrics, but they require analytical approaches that are not administratively and economically realistic for many system managers. Instead, an understanding of more nuanced aspects of resilience, such as panarchy, latitude, or precariousness, can then be built on the baseline these terms afford.

Finally, stakeholders are the ones who will ultimately decide whether to assess the resilience of their systems, and this hinges on their understanding of the value of these concepts. In many cases, it is best for system stakeholders to apply concepts in order to learn them, as opposed to learning concepts in order to apply them. The availability of user-friendly terms for straightforward system resilience analyses best suits this objective.

Taken together, these attributes are intended to provide the terms that are necessary and sufficient to describe the resilience posture of any system. Care was taken to develop terms that reflect system resilience rather than a desired end-state or system output. While there is significant overlap between the attributes, each term has been found adequately unique to stand as a separate trait. Note that certain attributes will play a more prominent role than others in any given situation, but the entirety of the list is intended to provide a firm foundation for assessing and managing resilience.

**TABLE 1:** Resilience Attributes.

| Stability Category | |
| --- | --- |
| Single Points of Failure | Singular features or aspects of the system, the absence or failure of which will cause the entire system to fail. |
| Pathways for Controlled Reductions in Function | Whether the functionality of a system, operation, or capability can be reduced in a manner that avoids the overwhelming effects of an unconstrained failure. |
| Resistance | The insensitivity of the system to stresses of a given size, duration, or character. |
| Balance | The degree to which a system is not skewed toward one strength at the expense of others. |
| Dispersion | The degree to which the system is distributed over space and time. |
| Adaptive Capacity Category | |
| Response Diversity | The variety and disparity of steps, measures, and functions by which an operation can carry out a task or achieve a mission. |
| Collaborative Capacity | The capacity to act through coordinated engagement. |
| Connectivity | How readily resources and information can be exchanged to ensure continued functionality. |
| Abundance/Reserves | The on-hand resource stores (capital) upon which a system can rely when responding to stress. |
| Learning Capacity | The ability to acquire, through training, experience, or observation, the knowledge, skills, and capabilities needed to ensure system functionality. |
| Readiness Category | |
| Situational Awareness | How well system, component, and functional capabilities are monitored. How readily emerging stresses or failures can be detected. |
| Simplicity/Understandability | How well system functions and capabilities can be understood. |
| Preparedness | The level of preparation in plans, procedures, personnel, and equipment for responding to system perturbations. |
| False Subsidies | Whether inputs, outputs, or internal processes receive incentives disproportionate or unrelated to their value. |
| Autonomy | A system manager's authority to select and employ alternate actions, configurations, and components in response to stress. |
| Enabling Traits | |
| Leadership and Initiative | The ability to motivate, mobilize, and provide direction in response to disruptions, as well as the ability to assume responsibility and act. |

The resilience attributes and their definitions are provided in Table 1. They are grouped into categories of Stability, Adaptive Capacity, and Readiness.

The attributes are sorted into the three categories (Table 2) to provide an easy cognitive basis for organizing the resilience attributes. It should be noted, however, that the attributes do not arise from the categories, but the categories instead arise from a convenient grouping of the attributes. Each of the attributes in fact features some degree of all three categories.

**TABLE 2:** Categories of Resilience Attributes.

| Single Points of Failure | Response Diversity | Situational Awareness |
|---|---|---|
| Controllable Degradation | Collaborative Capacity | Simplicity/Understandability |
| Resistance | Connectivity | Preparedness |
| Balance | Abundance/Reserves | False Subsidies |
| Dispersion | Learning Capacity | Autonomy |
| | Leadership and Initiative | |
| Stability | Adaptive Capacity | Readiness |

*Notes: Stability: The degree to which a system can continue to function if inputs, controls, or conditions are disrupted. It is a reflection of how minor a perturbation is capable of rendering the system inoperable or degraded; the types of perturbation to which the system is especially vulnerable; whether the system can "ignore" certain stresses; and the degree to which the system can be altered by surprise. Adaptive Capacity: The ability of a system to reorganize and reconfigure as needed to cope with disturbances without losing functional capacity and system identity. It reflects an array of response options and the ability to learn, collaborate, adapt, and create new strategies to ensure continued functionality. Readiness: How quickly a system can respond to changing conditions. It is affected by the physical, organizational, social, psychological, or other barriers, internal or external, that might impede timely response. Readiness is a measure of responsiveness; its converse is entanglement, a measure of the forces impeding responsiveness.*

The categories do, however, offer a useful functional construct. Put simply, managers need to know if the system as currently structured and resourced can survive a challenge (Stability), have the ability and options to respond if necessary (Adaptive Capacity), and understand if there are factors that help or hinder that response (Readiness).

### 3.2.4 ENABLING TRAITS

Complementing the three categories of resilience attributes, strong leadership and initiative are important factors for achieving resilience. Leadership is necessary to motivate, mobilize, and provide direction in response to disruptions, as is initiative to assume responsibility and act. These enabling traits transcend the categories depicted above and are foundational to the development and maintenance of all resilience attributes.

## 3.3 DESCRIPTIONS AND TARGETING QUERIES

The following are descriptions of the resilience attributes, which include capabilities that can be developed and honed to enhance system resilience; capacities that can be accrued or drawn upon to enhance resilience; and conditions that can be recognized and changed to a more resilient system configuration.

Each attribute is presented with associated "targeting queries" (assessed by the authors in [33] (p. 36–46)) to further assist in understanding the intention of each attribute and to support development of metrics for assessing and managing system resilience. (The term "targeting" is used to connote "aiming" for specific parameters within the system that manifest the attributes of resilience (or lack thereof) and can perform as efficient indicators.) The resilience attributes can be used widely and across systems; since effective metrics should be tailored to the specific system being managed, the targeting queries provide a means to bridge the gap and facilitate delineation of well-focused metrics.

Following each attribute description and its targeting queries below, simple examples are offered that depict tangible factors of potential interest in a resilience assessment in the context of two readily understandable systems: a small manufacturing company and a family farm. These systems perform based on dynamic interaction with their environment, but are simple enough to facilitate understanding.

### 3.3.1 STABILITY

Stability refers to the inherent ability of a system, as currently structured and functioning, to remain unaffected or minimally affected by disruptive forces. The concept of stability is well supported by the literature, including the concepts of latitude, resistance, and precariousness described by Walker, et al. [14].

### 3.3.1.1 STABILITY ATTRIBUTE: SINGLE POINTS OF FAILURE

- Singular features or aspects of the system, the absence or failure of which will cause the entire system to fail (See [22]).

Single Points of Failure can include physical, human, administrative, and other factors. They develop when a system is overly reliant on certain resources or capabilities that, if lost either temporarily or permanently, can threaten functionality. Systems may fail catastrophically, all at once. More commonly, however, specific stresses will challenge singular critical weaknesses in a system. Even seemingly robust systems have single points of failure, although the circumstances inducing emergence may be relatively rare.

**Targeting Queries:**
- On what physical (components, resources, linkages), human (manpower, skills, leadership, cultural, psychological, political), and administrative (organizational, legal, regulatory) factors does the system depend?
- Are there other critical system dependencies, including simultaneous or sequential functions, external systems, communications, or controls?
- Do single points of failure emerge only after a certain period of time? How well known are the time delays?
- Do single points of failure emerge without warning, or are there forewarnings or other indices? How well known are the warning signs?
- Are all who are affected by the system aware of the single points of failure?

**Examples:**

The success of the Smith Family Farm (henceforth the Farm) is vulnerable to insect, fungal, or bacterial pests that may damage crops. To diminish this threat, it cultivates several crops that are not likely to fail simultaneously.

Widget Corporation (henceforth Widget), an automotive parts manufacturer, is vulnerable because it requires debt to finance operations, and it can only get operating credit from one bank.

## 3.3.1.2 STABILITY ATTRIBUTE: PATHWAYS FOR CONTROLLED REDUCTIONS IN FUNCTION

- Whether the functionality of a system, operation, or capability can be reduced in a manner that avoids the overwhelming effects of an unconstrained failure. (As derived from Lovins' concept of stability [22].)

A system might not be able to retain its full function and identity beyond a certain duration or degree of external stress, but it can "fail gracefully" if it can maintain sufficient functionality long enough to engage compensatory measures or mitigation responses. A function's uncontrollable collapse is mitigated by enhancing Situational Awareness and by developing Response Diversity (specifically redundancy and substitutability) and Preparedness.

**Targeting Queries:**
- Is there sufficient information available to assess the emergence of system failure modes and to monitor controlled reductions in function?
- Are there methods for controlling a reduction in system function? Can problems be isolated and constrained to limit decline in system functionality?
- Are personnel trained in how to monitor conditions and adjust system components to control system degradation?
- Can a reduction in function be initiated before the onset of an unconstrained failure?
- Can controlled reductions in function be automated? If they already are automated, are they sufficient to handle all possible scenarios? Can manual controls override the automation if necessary?

**Examples:**

The Farm is able to fallow fields during drought in order to concentrate available water for the remaining crops to ensure a harvest and income.

When an influenza outbreak quickly depletes Widget's workforce, remaining cross-trained personnel are able to operate production lines at a reduced rate to turn out product.

## 3.3.1.3 STABILITY ATTRIBUTE: RESISTANCE

- The insensitivity of the system to stresses of a given size, duration, or character (See [14,34]).

Different systems, both static and dynamic, possess varying degrees of Resistance to different stressors; some systems are unaffected by stressors that may disrupt other systems. For some stressors, highly dynamic systems may be more resistant, while for other types of stressors, static systems may be more resistant (e.g., a willow tree versus an oak, depending on whether they are facing strong, gusty winds or physical assaults to their trunks).

**Targeting Queries:**
- Does the system have a history of being relatively unaffected by certain types of stresses?
- Can the system endure certain challenges for a known period or with a minimum of additional resources?
- Are there specific conditions under which the system is resistant to challenges and others under which it is more vulnerable or brittle? Can they be determined through analysis?
- Does the system indicate when its inherent resistance will be surpassed and failure will begin?

**Examples:**

The Farm practices no-till cultivation and plants cover crops, preserving the deep loamy soils from erosion.

Because Widget's process for parts making can accommodate materials of highly varied quality, it is not readily affected by changes in material suppliers and variation in the quality of raw materials.

## 3.3.1.4 STABILITY ATTRIBUTE: BALANCE

- The degree to which a system is not skewed toward one strength at the expense of others (See [12,14,34]).

A system is out of Balance when some inputs, controls, processes, or outputs change disproportionately to the rest of the system, thus weakening the system over time or creating vulnerability to certain types of stress. A system retains balance when it does not sacrifice certain strengths in favor of optimizing others. False Subsidies can skew a system out of Balance. Situational Awareness and Preparedness may be able to bring a system into Balance if sufficient Reserves are available.

**Targeting Queries:**
- Is the system skewed to a particular strength? If so, how is it skewed, and have other system attributes been sacrificed to achieve that strength?
- How well does the system handle a wide variety of missions and challenges?
- Is there a history of system performance faltering when faced with missions or stresses that differ greatly from the norm?
- Is the system subsidized to favor certain features, attributes, or capabilities?
- Can the system be tested to identify imbalances?

**Examples:**
The Farm has crops that ripen at different times over the course of the year and a timber lot from which to harvest in cold weather, spreading out its workload throughout the year.

Widget maintains complementary output from in-house and out-sourced production lines, has personnel skilled at multiple production tasks, and produces parts for more than one brand of automobile. In addition, when Widget invests capital in new equipment, it also establishes commensurate budgets for maintenance and training for that new equipment.

## 3.3.1.5 STABILITY ATTRIBUTE: DISPERSION

- The degree to which the system is distributed over space and time (See [22]).

Dispersion provides separation from systemic stressors. As a system evolves over time, dispersion may build resilience by fostering independent development of processes and capabilities, disparate strategies for responding to stress, and novel responses to external influences. Dispersion can be employed to reduce Connectivity and ensure that degradation in one part of a system does not disrupt the entire system. It also supports the development of Autonomy.

**Targeting Queries:**
- How is the system distributed (e.g., distance, time, physical barriers, technical separation, administrative or other organizational division, cultural or social separation, etc.)?
- Is separation sufficient to prevent the spread of systemic stresses?
- Are disparate elements free to act autonomously?
- Does the distribution drain resources or slow responses to challenges?

**Examples:**
The Farm is not physically dispersed, but concentrated in one valley. However the Farm does take advantage of temporal dispersion, as discussed in the previous example regarding the timing of harvests throughout the year.

Widget has two warehouses: a port facility to facilitate shipping and an inland facility sheltered from coastal storms.

## 3.3.2 ADAPTIVE CAPACITY

Adaptive Capacity is the ability of a system to reorganize and reconfigure as needed to cope with disturbances. Adaptive Capacity in ecological systems is "related to genetic diversity, biological diversity, and the heterogeneity of landscape mosaics," whereas, in social systems, it is "the existence of institutions and networks that learn and store knowledge and experience, create flexibility in problem solving and balance power among interest groups" [35].

## 3.3.2.1 ADAPTIVE CAPACITY ATTRIBUTE: RESPONSE DIVERSITY

- The variety and disparity of steps, measures, and functions by which an operation can carry out a task or achieve a mission. (As derived from [1,3,4,5,22,28,36,37,38]. See also [12,31,39]. See [13] for discussion of functional diversity.)

Response Diversity refers to the number and variety of options available to achieve a mission or task. It involves all aspects of a system—human-built and organic, subsystems and components, manpower and skill sets, institutional (administrative, managerial, legal, social), formal and casual. The ability to employ alternative components, features, skills, and strategies to accomplish an intended function can enhance a system's ability to withstand stresses. Variety enables system managers to select operational modes and capabilities that are either unaffected by perturbations or able to spread the force of the disturbance over multiple system facets, allowing the system to continue to function as intended. Similarly, response diversity ensures that the system can engage a range of responses to a variety of disturbances. Response Diversity includes such concepts as substitutability and redundancy. (Substitutability is a function of whether substitutable capabilities are flexible enough to work for a range of missions, and whether they can be engaged at different scales of application. It is affected by the amount of effort required to swap out one capability with another, and by whether, and to what extent, any capabilities might be degraded by a substitution. Subsets of substitutability are modularity and standardization, which address how readily system components and features can be exchanged.)

Development of response diversity is needed to counter the typical investments in efficiency and optimization that erode redundancy and flexibility within systems [40].

**Targeting Queries:**
- How easily can a mission, task, or function be accomplished in different ways or with different resources? How readily can this be done under stressed conditions?

- How many, how varied, and how well known are the options to accomplish a task?
- Are all aspects, components, features, and functions of the system covered?
- To what degree can substitute or redundant capabilities, components, sub-systems, controls, resources, skill sets, or features be combined, modified, or directly employed?
- At what cost to the system—immediately or over time—are substitutes employed?
- What burdens are placed on the system to maintain flexibility through redundancies and alternatives? Does the presence of redundancies foster complacency?
- How easily can response flexibility be incorporated into the system? Can changes in rules foster more creative responses to stressors?

**Examples:**

If the Farm's primary equipment for planting, cultivation, or harvest tasks are inoperable, the Farm employs a variety of back-ups (older equipment and alternate methods) to complete the tasks.

Widget employs the newest production technologies, but also maintains older machines in case the new equipment fails and replacement parts are not readily available. Additionally, while its predominant source of energy is the electric grid, it also maintains a natural gas-powered generator of sufficient size to keep operations going during power outages.

### 3.3.2.2 ADAPTIVE CAPACITY ATTRIBUTE: COLLABORATIVE CAPACITY

- The capacity to act through coordinated engagement. (As derived from [3,5,12,28,30,32,39]. See also [11].)

Collaborative Capacity refers to the potential of system managers to work cooperatively to ensure system function. It involves engaging linkages—relationships, authorities or permissions, and roles—in a timely and flexible manner to ensure system functionality. It also requires trust and shared understanding of the objectives of the collaboration. Collaborative capacity enables system managers to enlist (or provide) capabilities that would be too burdensome for any single actor to maintain.

**Targeting Queries:**
- Do personnel know others within the system with whom they can act, and how to make that coordination happen effectively?
- Can linkages be established and utilized in a timely manner?
- Do personnel recognize when to enlist others in collaboration? Is this personality-specific? Can it be instilled through training?
- Are personnel adequately motivated, and do they have the time, resources, and skills needed to collaborate?
- Can it be discerned when the benefits of collaboration outweigh the costs?

**Examples:**

The family that operates the Farm is close-knit, and they plan and work as a cohesive team.

Widget has good relationships with its material suppliers, who are willing to adjust their supply rates to meet Widget's needs if a surge or sudden drop in production is necessary. Widget's managers and blue-collar workers have a record of working together to solve design and manufacturing problems.

## 3.3.2.3 ADAPTIVE CAPACITY ATTRIBUTE: CONNECTIVITY

- How readily resources and information can be exchanged to ensure continued functionality. (See [11,12,39]. See also [13] for discussion of strong and weak interactions. See [41] for discussion of how inadequate or excessive connectivity may diminish system resilience.)

Connectivity refers to how readily a system can exchange resources and information with its environment and other systems to ensure continued function in the face of existential challenges. It includes cross-scale linkages with systems at larger and smaller scales, i.e., with systems of which it is a part hierarchically and with its own subsystems, as well as any other systems accessible through existing or improvised links. Connectivity confers resilience by the response flexibility and Situational Awareness it enables, allowing systems to proactively alter their readiness posture in anticipation of looming challenges, or to rapidly exploit information and resources in response to surprises. While Connectivity may help avoid system failure by allowing stresses to be spread over several systems, it may also hasten an even larger collapse as the demands of one failing system can overwhelm others from which it draws support. As such, it may be desirable to have connective links that can be decoupled, isolating threatening disturbances and preventing larger failures. Connectivity involves feedback loops that signal how system activities affect connected systems and subsystems; loose feedback loops may reduce resilience by slowing system response to disturbance or masking system affects upon sub- or adjacent systems.

**Targeting Queries:**
- Where, when, and how are information and/or resources exchanged?
- Are the pathways and links for that exchange known? How well maintained are they? How effective are they? How often are they used?
- Are pathways of connectivity personality-dependent, or can they be accessed and employed by anybody when necessary?
- How does connectivity support response flexibility and situational awareness? How is it enabled by situational awareness?
- What resources are allocated to maintain connectivity?
- Can connectivity pathways and links be severed when necessary to prevent the spread of problems?
- Do bureaucratic requirements slow or prevent action?

**Examples:**
Due to the independent personality of the farmer, the Farm does not take advantage of opportunities such as the Farmer's Co-operative and University Extension service.

Widget is in constant communication with automobile makers to stay abreast of changes that will affect the design of parts. However, it is receiving—and reacting to—information before those changes have been confirmed, and has had to restrict its communications to the exchange of final, approved new designs.

*3.3.2.4 ADAPTIVE CAPACITY ATTRIBUTE: ABUNDANCE/RESERVES*

- The on-hand resource stores (capital) upon which a system can rely when responding to stress. (See [33,39]. See also [3,9,32].)

Abundance/Reserves refers to the surplus within a system of various forms of capital (natural, economic, social, built, etc.) that enable it to prolong functionality in the face of a stressor. Abundance provides the resources that support variety, redundancy, preparedness, and other factors contributing to and prolonging a system's resilience.

**Targeting Queries:**
- What resources does the system maintain for immediate engagement when stressed?
- Are the system's reserves monitored and their limits known?
- Is the system made brittle, vulnerable, or less stable when it employs its resources?
- Are there conditions under which the system's resources are rendered unavailable?
- How is a sufficiency of resources determined?

**Examples:**
The Farm is vulnerable due to its lack of cash reserves and reliance on credit to finance planting and equipment repair expenditures.

Widget practices just-in-time manufacturing and keeps only a very small supply of excess raw materials. As a result, it must scale back production when faced with delivery problems that constrain availability of materials.

## 3.3.2.5 ADAPTIVE CAPACITY ATTRIBUTE: LEARNING CAPACITY

- The ability to acquire, through training, experience, or observation, the knowledge, skills, and capabilities needed to ensure system functionality. (ibid. See also [1,23,25,26,28,31,39,40,42].)

Learning Capacity involves the ability to draw upon and combine different types of knowledge to support system readiness for, and responses to, disturbances. It may be a trait of individuals as well as organizations. It can be obtained experientially, in a classroom, or through electronic communications, supported by structured and intentional efforts as well as by unplanned and circumstantial events.

**Targeting Queries:**
- Is there an individual and organizational culture of learning?
- Are there active adaptive management and lessons-learned programs in place?
- Have personnel received expected training?
- Is there institutional support for increased education/training?
- Is the system sufficiently manned to allow personnel the time needed for training?

**Examples:**

The farmer plans which crops to plant in which locations based on knowledge of what has succeeded on the past and is hesitant to experiment with new approaches. This practice exhibits passive, rather than active adaptive learning.

Widget pays to keep its employees' skills current via on-the-job-training and formal course work. As a result, the company is able to employ new production technologies and practices that enhance Widget's competitiveness.

### 3.3.3 READINESS

Readiness is a measure of responsiveness; its converse is entanglement, a measure of the forces impeding responsiveness. While Readiness is affected by physical traits and components, it is more prominently driven by organizational, administrative, legal, social, and related institutional factors. Factors contributing to readiness may arise as a system evolves to a given functional state, and altering those factors may quickly challenge that functionality. Analyses in this realm may reveal unexpected and persistent impediments to a system's operational effectiveness.

#### 3.3.3.1 READINESS ATTRIBUTE: SITUATIONAL AWARENESS

- How well system, component, and functional capabilities are monitored.
- How readily emerging stresses or failures can be detected. (See [31,36,37,38,39].)

Situational Awareness refers to how well the status of a system is monitored. It is a measure of how readily emerging changes and failures can be detected, recognized, and acted upon to minimize adverse effects. Situational awareness includes recognition of a system's potential tipping points and of possible means to avoid passing them; this includes an awareness of how system features and components afford opportunities or vulnerabilities in the face of challenges. It is a measure of the ability to recognize critical dependencies. Situational Awareness relies on the availability and timeliness of accurate and useful information, including sufficiently frequent updates.

Situational awareness is affected by the presence of formal and informal structures (e.g., training and mentoring programs, or regular checks of the system and its environment) that enhance individual and institutional learning. It reflects the ability to recognize critical dependencies and relies on connecting internally among subsystems and externally to other systems.

**Targeting Queries:**
- Does monitoring take place to detect and identify stresses?
- How timely and understandable is the information provided?
- How comprehensive is the information about the system and its environment? Conversely, how well known are the information gaps?
- Can queries yield additional information?
- How well are personnel trained in knowledge of the overall system; in the use of system monitoring technology; and how to capitalize on advantages designed into the system?

**Examples:**

The farmer is situationally aware because he is in the fields every day and monitors weather projections and commodity markets on a regular basis.

By monitoring developments in the automotive market and changes in consumer preferences, Widget is able to anticipate new production needs and posture itself to employ the equipment and materials that will satisfy market demands.

## 3.3.3.2 READINESS ATTRIBUTE: SIMPLICITY/ UNDERSTANDABILITY

- How well system functions and capabilities can be understood. (As derived from Lovins' concept of accessibility [22].)

Simplicity/Understandability is the degree to which a function or capability is readily understood by those it affects. This does not mean that a system must be simple to be understood; it refers instead to how well it is comprehended by people acting within the system. It also encompasses an understanding of a system's hierarchical connections, as well as the environment in which it exists and on which it depends.

Simplicity/Understandability can be enhanced through technology (sensors and visual aids that explain system status and function), techniques (procedures that break down the function into easily understood steps), and strategies (the culling of the excess and superfluous). Moreover, familiarity with new items, actions, or rules is gained with daily exposure, while infrequent engagement may foster an unsure grasp that could greatly diminish system function during stressful periods.

**Targeting Queries:**
- How, and to what degree, is system understanding achieved and maintained?
- Can the complete system be understood by combining partial information from multiple sources?
- How is system understanding shared or transferred? How readily can a newcomer understand the system?
- Can richer information about the system be obtained? How?

**Examples:**
The Farm's operations are uncomplicated and environmental factors are well understood by the farmer since his family has farmed this location for three generations.

Due to changes in key management, new Widget corporate personnel do not fully understand the company's underlying financial arrangements. Similarly, with the retirement of long-term plant employees and the incorporation of highly automated machines, leadership does not fully comprehend the production line's capabilities and constraints. However, the plants do have adequate records, standing operating procedures, and manuals that can be used to rectify these situations.

### 3.3.3.3 READINESS ATTRIBUTE: PREPAREDNESS

- The level of preparation in plans, procedures, personnel, and equipment for responding to system perturbations. (See [34].)

Preparedness refers to the existence of plans and procedures by which a system can respond to perturbations and stressors. It addresses whether contingencies have been considered for expected disturbances as well as disturbances for which little consideration would normally be given but for which the system is particularly vulnerable. Preparedness may involve formal plans that are tested, regularly exercised, and kept current; informal plans that are developed on an impromptu basis; or some combination thereof. Such plans would address how to reinforce or substitute existing subsystems and capabilities, adapt new strategies, and otherwise mitigate lost or threatened functionality.

**Targeting Queries:**
- Do response plans and procedures exist? Are they formal or informal? Are they flexible? How readily can they be modified for unforeseen circumstances?
- How accessible are plans and procedures? Do those affected know how and where to access them? Are they dependent on specific personnel?
- Are the plans well maintained, frequently updated, and tied to training and exercises?
- How readily implemented are plans and procedures?
- Are personnel well prepared and aware of threats?
- Is equipment well maintained?

**Examples:**
The farmer does not have written plans for emergencies, but he and his family are well practiced in responding to unexpected conditions and emergencies.

Because Widget has suffered during previous material and manpower shortages, it has developed contingency plans in case those problems re-emerge. It has not updated or practiced those procedures, however.

## 3.3.3.4 READINESS ATTRIBUTE: FALSE SUBSIDIES

- Whether inputs, outputs, or internal processes receive incentives dispropor-
  tionate or unrelated to their value. (See [30] (p. 26), [39].)

False Subsidies refer to whether a system's capabilities or features receive support that exceeds, or is unrelated to, the benefit it provides. These
skewing incentives may come in many forms, including financial, material, organizational, legal, social, and cultural. They may be formal or informal, sought or imposed. False Subsidies influence a system to function
in a manner different than it normally would, given the natural conditions,
the natural range of variability and suite of inputs and outputs, in which the
system evolved. The need for subsidies may emerge unintentionally and
undetected as a system evolves to perform a given function, hidden in the
assumptions underlying the system's formation.

**Targeting Queries:**
- Are any false subsidies known or readily identified?
- How readily can any subsidies be discontinued, either temporarily or per-
  manently?
- Who controls the false subsidies, and how engaged is that controlling entity
  in the function or purpose of the system?
- Can false subsidies be converted to assets that do not skew the system?

**Examples:**
The Farm receives crop insurance and commodity subsidies. These guar-
antees influence the farmer to plant a higher proportion of high risk – high
reward crops than he would if unsubsidized.

The price of certain mined minerals has been kept artificially low
due to government subsidies. Widget has structured its entire production
capability around materials that are based on those minerals. If national
sentiment about the environmental hazards of mineral mining leads government officials to end subsidies, Widget could face drastic increases
in the cost of materials and of retrofitting its production line to use new,
different feedstock.

## 3.3.3.5 READINESS ATTRIBUTE: AUTONOMY

• A system manager's authority to select and employ alternate actions, configurations, and components in response to stress. (Per [36,37,38].)

Autonomy refers to the degree to which an organization, operation, or function can self-select alternate actions, configurations, and strategies to achieve the specific mission or function—essentially, control over its destiny. Autonomy enables an actor to self-organize and choose the timing, order, and priority of actions deemed appropriate for a given circumstance to avoid systemic failure. It allows an actor to select or establish the relationships necessary to function, and to loosen, tighten, or otherwise change the nature of those linkages as necessary. It also allows the actor to make trade-offs that ensure continued system functionality. A system that requires permission in how and when to act may encounter costly delays and receive instructions from those not close enough to fully understand the nature of the problem. Systems defined by Command and Control "pathologies" often struggle to express autonomy [1,4].

**Targeting Queries:**
• Is the hierarchy of authority or power structure known? Is it necessary, and under what circumstances can it be bypassed?
• Can autonomy be exercised on a situational basis, e.g., in proportion to the stressor, for specific stresses or system features, or on a time-limited basis?
• Are personnel trained to handle autonomous decision-making?
• Does the right or authorization to act autonomously include the ability to negotiate and coordinate with other parties?
• Do parties within and outside of a system recognize others' authority to act autonomously?

**Examples:**
The Farm operates autonomously within the constraints of federal and state regulations, debt structure, and commodity contracts.

Widget's management approach is highly hierarchical, so employees must receive permission before making even simple changes in how they accomplish their work. When production problems emerge, much time is wasted awaiting authorization to make necessary fixes.

## 3.3.4 ENABLING TRAITS

Leadership and initiative connote sentient agency within an SES; complementing the three categories of resilience attributes, they are important factors for achieving resilience in human-managed systems. Strong leadership is necessary to motivate, mobilize, and provide direction in response to disruptions; it is underscored by initiative to assume responsibility and act. These enabling traits transcend the categories depicted above and are foundational to the development and maintenance of all resilience attributes.

Leadership and initiative respond to changing conditions and to the flow of information and resources, and are informed by knowledge of the system's structure, functions, culture, politics, and history. Leaders select from possible responses and create new options in reaction to shocks and disturbances, manifesting as the adaptability and flexibility needed for resilience. Leadership ensures accountability. Leadership tempers initiative, using judgment to determine when to effectively engage a problem and when certain 'normalizing' actions, which are often based on cultural and system design priorities, might create greater system fragility; in such cases, patience and the development of alternate courses of action may be necessary. Finally, leadership and initiative involve knowing how to leverage actions to greatest effect and how to moderate any actions so as to achieve the most desirable outcomes [6,26,27,29].

**Targeting Queries:**
- Are there actors within the system who fill a leadership role? Are these actors strong leaders?
- Are leaders apparent and agreed upon? Do leaders recognize themselves as such, either on a formal, ad hoc, or implicit basis? Do outside parties, as well as other actors in a system, recognize leaders as well? Do the leaders possess the authority to affect changes and negotiate with governmental agencies and outside actors?
- Do the leaders have enough history within the system to be knowledgeable about system conditions, vulnerabilities, and internal and external threats?
- Do system rules, explicit or implicit, present incentives or obstacles to stakeholder initiative in the face of system challenges?
- Are the leaders comfortable with adaptive management, or do they seek less flexible approaches?

- Do system design parameters and constraints engage and support the performance of leaders when shocks and disturbances challenge initial operating assumptions and change operating conditions?

**Examples:**

Leadership of the Farm is very clear, with the farmer having authority over all operations. Outside parties recognize the farmer's role as leader of the Farm.

Widget's management approach is reactive and lacks initiative. Employees complain about not having the delegated authority to anticipate needs and adjust processes when needed.

## 3.4 DISCUSSION AND CONCLUSIONS

The resilience attributes were developed to characterize human-managed systems, where policies and practices can greatly alter system dynamics, and they have been refined through a dozen iterations and tested in a resilience assessment that is the subject of a forthcoming journal article. Development of these resilience attributes has been a synthesis of systems engineering terminology and the holistic ecological resilience theory for SESs that has emerged since the 1970s.

Some researchers have developed resilience rating indices based on evaluation variables deemed essential for the continued function of all components making up a system (for example, "does a community have a certified flood plain manager" or "will a certain percentage of road be passable within one week of a major storm"). Cutter et al. developed a composite index of disaster resilience indicators for communities. The index is designed to assess current programs and policies for their likely effectiveness in improving disaster resilience [43]. This approach differs from the one described herein in that it does not appear to have been derived from a set of organizing resilience principles, but rather organizes around five sectors of society: social, economic, institutional, infrastructure, and community capital. For each sector, a series of variables have been developed that are presented as universally applicable across systems, but the underlying basis within resilience theory for each variable is not well defined.

A similar approach is taken in the Coastal Community Resilience Index (CCRI), a self-assessment tool to aid community leaders in determining how their community will function following disaster [44]. This tool evaluates critical infrastructure and facilities, transportation, community plans and agreements, mitigation measures, and business plans without examining the underlying principles or resilience attributes of the system being assessed—perhaps appropriate for a lay-person self-assessment, but it may be insufficient for comparison across systems, or for deeper analysis necessary to predict system performance during various categories of disruption. Nor does the CCRI have a process for examining how the individual components within each category may affect the other components within that category or in other categories. Synergies and confounding affects seem to be ignored.

Cumming et al. developed an exploratory framework for empirically measuring resilience that focused on four system components necessary to support system identity—their focal points for measuring resilience: structural components that make up the system; functional relationships that link components; innovation variables that relate to development of novel solutions and responses to change; and continuity variables that maintain identity through space and time [41]. The approach presented in this article is similar in that it presents a means for assessing the SES for attributes that can be managed to maintain system identity.

The set of resilience attributes presented in this article reflects a synthesis of many such assessment approaches, indices, and lists developed over several decades, fashioned into a whole for the purpose of providing a comprehensive resource for those professionals tasked with managing systems. These attributes, in concert with the above "targeting queries, enable managers to develop system-specific resilience metrics, which they can incorporate into their plans and programs and make their SESs more robust to disruption. This approach ties measured system parameters directly to those attributes that connote resilience within that system, thus informing managers of how policies and management interventions may relate to building or diminishing system resilience.

As noted earlier, certain variables will take on greater significance than others for a given situation, but taken together the entire collection of attributes is intended to provide a firm basis for beginning the resilience assessment process. Moreover, changes that foster improvements in some areas of resilience may reduce it in others; trade-offs will likely be necessary.

## 3.4.1 LIMITATIONS

SES stakeholders are usually able to contend with challenges about which they are aware and concerned. The range of resilience attributes provided herein may foster greater awareness of potential concerns not previously considered. However, the attributes, as posited, focus on a steady-state, moment-in-time view of the system. In complex SESs, shocks and disturbances create spatial, temporal, and functional reactions in the system; the extent, trend, or dynamics of these reactions may exceed what the resilience attributes-based snapshot is able to capture. Resilience concepts that address such dynamics include panarchy, with nestedness and cross-scale interactions; thresholds; transformability; latitude and precariousness (within a basin of stability); and acting to manage the slow and fast variables of internal and external threats and dynamics. The resilience attributes also do not consider how management interventions may lead to second- and third-order effects in adaptive and often stochastic, unpredictable systems.

In theoretical terms, it is difficult to characterize the stability domain within which a system resides, including its depth and shape, and the system's location relative to the walls and its ease or difficulty of movement, without knowing the full range of possible stressors and how the system may react to them. Similarly, it is difficult to know where a system resides within the adaptive cycle without knowledge of the nature (type, timing, and strength) of a disturbance as well as the system's status and possible responses to the disturbance. The resilience attributes do not reveal the dynamics driving the system, and require monitoring to discern response trends (directions and magnitudes), shifts indicating the passing of thresholds, and the triggering of novel response dynamics. The predictive value of the resilience attributes for scenario analyses are thus limited by stakeholder understanding of underlying system dynamics, which, as noted, may be constrained by experience, expectations, awareness, and a capacity and willingness to explore.

## 3.4.2 BENEFITS

As noted earlier, all resilience analysis strategies must contend with the challenges associated with assessing and managing the resilience of complex, adaptive systems, and all strategies must, at some point, characterize the systems in question in a consistent manner that supports deeper analysis. This includes modeling, simulation, adaptive management, and similar approaches. Recognizing the limitations mentioned above, there are distinct benefits that may be derived from the resilience attributes. They provide a common approach for characterizing the resilience of different systems that enables program managers and other stakeholders to aggregate resilience assessment results and chart progress toward goals over time and among disparate areas of assessment. (If the underlying metrics for resilience are incompatible, this is difficult to do in any meaningful way.) Additional benefits of a consistent approach include the following:

- Supports informed decision-making by providing a common basis for comparing and understanding resilience across systems.
- Supports identifying and forecasting potential problems and tipping points, instead of learning from failures or from a system on the edge of failure. The whole-system perspective reveals "least-harm solutions."
- Provides additional insights about the effects of policies and practices on system function in the face of unforeseen challenges.
- Provides a solid basis for developing baselines against which to measure trends and progress. This may entail formulating scenarios for qualitatively different future trajectories, then setting resilience metrics-based "waypoints" or "guideposts" in each scenario to provide early warning to managers that the system is evolving in a particular direction. It also supports the evaluation of investment and intervention options [45].
- Provides an additional line of evidence for managing facilities and environmental resources, and for making investment decisions by land use and regulatory agencies, federal installations, states and municipalities, and energy and water utilities, for example.
- Ties to Strategic Planning Objectives and Milestones. Generating quality metrics is typically the most challenging part of the strategic planning process. Using these resilience attributes and targeting queries to develop system-specific metrics can greatly improve strategic plans while incorporating resilience principles.
- Supports Adaptive Management. The resilience attributes lend themselves to assessing system thresholds and sensitivities within an active adaptive management program.

TABLE 3: Examples of Applications for Resilience Attributes and Metrics.

| Category | Application |
|---|---|
| Permanent Federal and State Installations, Park Lands and Reserves | Perform an assessment of critical infrastructure's resilience to climate change. |
| | Assess forests, range lands, and wetlands for resilience in the face of climate change. |
| | Incorporate resilience metrics into Natural Resource Management Plans, Cultural Resource Management Plans, and similar planning processes. |
| | Perform a facility and infrastructure resilience assessment in response to the 2013 Executive Order on Climate Change Adaption and Resilience. |
| Temporary Remote Installations | Use resilience attributes to support "war gaming" and other planning for various ways to configure and allocate resources for military contingency bases or scientific research outposts for enhanced resilience. |
| Military Operational Energy and Water | Perform a comparative analysis of resilience in operational energy and water alternatives (e.g., the resilience of energy and water supply and delivery systems as well as of the overall systems using those resources). |
| | Incorporate resilience as an additional line of evidence for making investment decisions. |
| Watershed Management | Incorporate resilience metrics into watershed management plans and water quality improvement initiatives. |
| International Development and Aid Organizations | Assess the resilience of fragile states (by sector and as a whole) and the resilience implications of various assistance initiatives. |
| Communities | Incorporate resilience metrics into community preparedness and emergency response plans. |
| | Assess the disaster preparedness of a community and tailor emergency response exercises toward the weakest sectors. |
| | Evaluate the resilience of rural agro-economies and develop the funding rationale for capacity building and development. |
| Critical Infrastructure Security and Resilience (Presidential Policy Directive-21) | Develop uniform analytical metrics and assessment methodology for Federal agencies' response to PPD-21. |

## 3.4.3 APPLICATIONS

The resilience attributes are intended for decision-support at multiple scales within an integrated system. This scalability enables the development of well-aligned organizational resilience strategies, policies, programs, and training. For example, State water managers might focus on building water resilience at several scales: first, by increasing redundant sources of supply for critical applications state-wide (involving the resilience attribute of *Response Diversity*), and then by exercising how to deal with electric power interruptions and water delivery interruptions regionally (*Preparedness* and *Pathways for Controlled Reductions in Function*).

Resilience may be increased immediately; consider, for example, a power utility. By developing additional electric grid interconnections and alert systems (*Connectivity*), managers increase their power networking options. They can also focus efforts to build resilience by conducting drills and exercises to test for *Single Points of Failure*, critical dependences, and the ability to substitute power types for key systems, e.g., testing regular and extended use of multiple fuels within multi-fuel platforms (*Response Diversity*).

Agricultural program managers may reconsider policies with an eye toward realigning or eliminating incentives (e.g., Federal or State tax credits for subsidized water import) that provide funding support but work against making recipients and the regional community more resilient (False Subsidies).

Specific areas for potential application of the resilience attributes, and suggestions for how they might be employed, are delineated in Table 3.

We advance the suite of resilience attributes described herein to aid SES managers in recognizing threats and vulnerabilities. The resilience attributes provide a basis for developing resilience metrics that support existing management plans and programs and bring a new perspective for prioritizing objectives, planning resource allocation, and defending investment decisions.

### 3.4.4 FUTURE RESEARCH

These attributes were tested in a 2013 assessment for the U.S. Army Rapid Equipping Force of the energy and water resilience of a representative combat outpost to a range of difficult shocks and perturbations. The study, which highlighted the resilience vulnerabilities of the combat outpost, provided a preliminary assessment of the applicability and effectiveness of the resilience attributes. While useful, that assessment would be greatly aided by further investigation into the value, strengths, and limitations of the resilience attributes. The following lines of inquiry are suggested:

- Are the terms sufficiently comprehensive and understandable? How easily can stakeholders employ them in creating useful metrics for system management? In addition, can they be used consistently by different users over time?
- How do the attributes relate to and support assessment of such concepts as panarchy, thresholds, transformability, latitude, precariousness, and managing slow and fast external variables? How might they be employed beyond a baseline analysis in an iterative approach to resilience assessment?
- Can the attributes alone characterize system resilience (and if so, to what degree)? Do they surpass simple metrics generated to evaluate infrastructure and emergency and backup functions (e.g., the community resilience index [44])? Are the attributes better suited for framing a deeper resilience analysis (e.g., of thresholds or panarchy)? Are deeper analyses any more reliable than the baseline snapshot offered by these resilience attributes? How can that be tested?
- To what applications do the resilience attributes best lend themselves? This may include vulnerability analyses and gap assessments; assessing the implications of system changes; tracking system conditions and trends over time; prioritizing functions for graceful degradation; and planning investments in infrastructure or other system components.
- How are the attributes best employed? This may include scenario analyses; metrics to guide strategic planning, monitoring programs, and adaptive management; and "Dashboard" depictions for routine monitoring.

In addition to the above research, it may also prove useful to explore the possibility and value of a short, field-expedient version of the resilience attributes. Such a tool would enable stakeholders engaged most closely with a system to contribute to an overall understanding of, and to enhance, the system's resilience posture.

## REFERENCES

1.  Holling, C.S. Resilience and stability of ecological systems. Annu. Rev. Ecol. Syst. 1973, 4, 1–23.
2.  Fiksel, J. Sustainability and resilience: Toward a systems approach. Sustain. Sci. Pract. Policy 2006, 2, 14–21.

3. Walker, B.; Salt, D. Resilience Thinking: Sustaining Ecosystems and People in a Changing World; Island Press: Washington, DC, USA, 2006; pp. 1–151.

4. Folke, C.; Carpenter, S.; Elmqvist, T.; Gunderson, L.; Holling, C.S.; Walker, B. Resilience and sustainable development: Building adaptive capacity in a world of transformations. Ambio 2002, 31, 437–440.

5. Walker, B.; Gunderson, L.; Kinzig, A.; Folke, C.; Carpenter, S.; Shultz, L. A handful of hueristics and some propositions for understanding resilience in social-ecological systems. Ecol. Soc. 2006, 11, 1–8.

6. Walker, B.; Salt, D. Resilience Practice: Building Capacity to Absorb Disturbance and Maintain Function; Island Press: Washington, DC, USA, 2012; pp. 85–100.

7. National Academy of Sciences (NAS). Disaster Resilience: A National Imperative; National Academy Press: Washington, DC, USA, 2012; pp. 1–31.

8. The Johnson Foundation at Wingspread. Building Resilient Utilities: How Water and Electric Utilities Can Co-Create Their Futures; The Johnson Foundation at Wingspread: Racine, WI, USA, 2013; pp. 2–3.

9. Resilience Alliance. Assessing Resilience in Social-Ecological Systems: Workbook for Practitioners (Revised Version 2). Available online: http://www.resalliance.org/workbook (accessed on 30 November 2014).

10. Folke, C.; Carpenter, S.; Walker, B.; Scheffer, M.; Elmqvist, T.; Gunderson, L.; Holling, C.S. Regime shifts, resilience, and biodiversity in ecosystem management. Annu. Rev. Ecol. Evol. Syst. 2004, 35, 557–581.

11. Walker, B.; Carpenter, S.; Anderies, J.; Abel, N.; Cumming, G.S.; Janssen, M.; Lebel, L.; Norberg, J.; Peterson, G.D.; Pritchard, R. Resilience management in social-ecological systems: A working hypothesis for a participatory approach. Conserv. Ecol. 2002, 6, 14.

12. Biggs, R.; Schluter, M.; Biggs, D.; Bohensky, E.L.; BurnSilver, S.; Cundill, G.; Dakos, V.; Daw, T.M.; Evans, L.S.; Kotschy, K.; et al. Toward principles for enhancing the resilience of ecosystem services. Annu. Rev. Environ. Resour. 2012, 37, 421–448.

13. Thrush, S.F.; Hewitt, J.E.; Dayton, P.K.; Coco, G.; Lohrer, A.M.; Norkko, A.; Norkko, J.; Chiantore, M. Forecasting the limits of resilience: Integrating empirical research with theory. Proc. R. Soc. 2009, 276, 3209–3217.

14. Walker, B.; Holling, C.S.; Carpenter, S.R.; Kinzig, A. Resilience, adaptability and transformability in social-ecological systems. Ecol. Soc. 2004, 9, 5.

15. Holling, C.S.; Gunderson, L.H. Resilience and adaptive cycles. In Panarchy; Gunderson, L.C., Holling, C.S., Eds.; Island Press: Washington, DC, USA, 2002; Chapter 2. pp. 25–62.

16. Gunderson, L.; Folke, C. Resilience 2011: Leading transformational change. Ecol. Soc. 2011, 16, 30.

17. Holling, C.S.; Gunderson, L.H.; Peterson, G.D. Sustainability and panarchies. In Panarchy; Gunderson, L.C., Holling, C.S., Eds.; Island Press: Washington, DC, USA, 2002; Chapter 3. pp. 63–102.

18. Kidd, R. Deputy assistant secretary of the army (energy & sustainability). In Proceedings of the Comments at the Environment, Energy Security & Sustainability (E2S2) Symposium & Exhibition, Panel on Energy Security Resilience. New Orleans, LA, USA, 23 May 2012.

19. Mitchell, A. Comments during briefing on "Managing Energy and Water Resilience". Washington, DC, USA, Unpublished work. 19 February 2014.
20. Nagel, A.M. Comments during briefing on "Managing Energy and Water Resilience". Washington, DC, USA, Unpublished work. 19 February 2014.
21. National Defense Industrial Association (NDIA). Top Issues 2013; NDIA: Arlington, VA, USA, 2013. Available online: http://www.ndia.org/Advocacy/PolicyPublicationsResources/Documents/TopIssues_2013_web.pdf (accessed on 30 November 2014).
22. Lovins, A.; Lovins, H. Brittle Power: Energy Strategy for National Security; Brick House Publishing Co., Inc.: Andover, MA, USA, 1982; pp. 1–333.
23. Lee, K. Compass and Gyroscope: Integrating Science and Politics for the Environment; Island Press: Washington, DC, USA, 1993; pp. 51–86.
24. Becker, C.D.; Ostrom, E. Human ecology and resource sustainability: The importance of institutional diversity. Annu. Rev. Ecol. Syst. 1995, 26, 113–133.
25. Folke, C.; Carpenter, S.R.; Walker, B.; Scheffer, M.; Chapin, T.; Rockström, J. Resilience thinking: Integrating resilience, adaptability and transformability. Ecol. Soc. 2010, 15, 20.
26. Ostrom, E. Understanding Institutional Diversity; Princeton University Press: Princeton, NJ, USA, 2005; pp. 1–48.
27. Ostrom, E. A general framework for analyzing sustainability of social-ecological systems. Science 2009, 325, 419–422. [PubMed]
28. Berkes, F. Understanding uncertainty and reducing vulnerability: Lessons from resilience thinking. Nat. Hazards 2007, 41, 283–295.
29. Olson, P.; Gunderson, L.H.; Carpenter, S.R.; Ryan, P.; Louis, L.; Folke, C.; Holling, C.S. Shooting the rapids: Navigating transitions to adaptive governance of social-ecological systems. Ecol. Soc. 2006, 11, 18.
30. Thomas, S.; Kerner, D. Defense Energy Resilience: Lessons from Ecology; Letort Paper. Strategic Studies Institute, United States Army War College: Carlisle, PA, USA, 2010; pp. 1–35.
31. Anderies, J.M.; Folke, C.; Walker, B.; Ostrom, E. Aligning key concepts for global change policy: Robustness, resilience, and sustainability. Ecol. Soc. 2013, 18, 8.
32. Stokols, D.; Lejano, R.P.; Hipp, J. Enhancing the resilience of human-environment systems: A social-ecological perspective. Ecol. Soc. 2013, 18, 7.
33. Army Rapid Equipping Force (Army REF). Final Report Energy Resilience Study: Resilience Assessment of Notional Combat Outpost; Contract SP0700-D-0301.16. Army Rapid Equipping Force (Army REF): Alexandria, VA, USA, 2013; pp. 1–76.
34. Thomas, S.; Kerner, D. Metrics for assessing resilience: A social-ecological systems perspective. In Proceedings of the Presentation to the Challenges of Natural Resource Economics and Policy National Forum on Socioeconomic Research in Coastal Systems, New Orleans, LA, USA, 25 March 2013.
35. Resilience Alliance. Adaptive Capacity. Available online: http://www.resalliance.org/index.php/adaptive_capacity (accessed on 30 November 2014).
36. Kerner, D.; Thomas, S. Efficiency and conservation not enough to achieve energy security. National Defense Magazine 2012, 1–3.
37. Kerner, D; Thomas, S. Understanding energy resilience. In Proceedings of the Presentation to the Energy Security Panel, Energy and Environmental Sustainability

Symposium, National Defense Industrial Association, New Orleans, LA, USA, 23 May 2012.

38. Kerner, D.; Thomas, S. Defense energy resilience: Assessment metrics. In Proceedings of the Presentation to the Energy Security Panel, Energy and Environmental Sustainability Symposium, National Defense Industrial Association, New Orleans, LA, USA, 23 May 2012.

39. Carpenter, S.R.; Arrow, K.J.; Barrett, S.; Biggs, R.; Brock, W.A.; Crepin, A.; Engstrom, G.; Folke, C.; Hughes, T.P.; Kautsky, N.; et al. General resilience to cope with extreme events. Sustainability 2012, 4, 3248–3259.

40. Park, J.; Seager, T.P.; Rao, P.S.C.; Convertino, M.; Linkov, I. Integrating risk and resilience approaches to catastrophe management in engineering systems. Risk Anal. 2013, 33, 356–367.

41. Cumming, G.S.; Barnes, G.; Perez, S.; Schmink, M.; Sieving, K.S.; Southworth, J.; Binford, M.; Holt, R.D.; Stickler, C.; van Holt, T. An exploratory framework for the empirical measurement of resilience. Ecosystems 2005, 8, 975–987.

42. Olsson, P.; Folke, C.; Berkes, F. Adaptive comanagement for building resilience in social-ecological systems. Environ. Manag. 2004, 34, 75–90.

43. Cutter, S.L.; Burton, C.G.; Emrich, C.T. Disaster resilience indicators for benchmarking baseline conditions. J. Homel. Secur. Emerg. Manag. 2010, 7, 1–22.

44. Sempier, T.T.; Swann, D.L.; Emmer, R.; Sempier, S.H.; Schneider, M. Coastal Community Resilience Index: A Community Self-Assessment. Available online: http://masgc.org./uploads/publications/364/08-014.pdf (accessed on 30 November 2014).

45. Thomas, S; Mouat, D. Alternative futures analysis as a complement to planning processes for the use of military land. Air Space Power J. 2011, 25, 100–109.

# PART II

# GENERAL RESILIENCE STRATEGIES

# CHAPTER 4

# Community Vitality: The Role of Community-Level Resilience Adaptation and Innovation in Sustainable Development

ANN DALE, CHRIS LING, AND LENORE NEWMAN

## 4.1 INTRODUCTION

The environmental movement, and the parallel in international policy development, has evolved significantly since its modern rise to prominence in the early 1960s with, amongst other events, the publication of *Silent Spring* [1] and the resulting activism and political resurgence that followed. Beginning as a grass-roots movement against very specific threats, the environmental movement evolved into an effort of international scope. The maturing UN policy agenda starting with the Intergovernmental Conference for Rational Use and Conservation of Biosphere in 1968 and culminating in the production of the Brundtland report "Our Common Future" [2] and Agenda 21 [3]. While these various international policy outcomes were the products of intense diplomacy, contained visionary ideas and concepts, and were grounded in significant scientific research, the outcome and impact of this effort was rather disappointing in some ways. The impact of humanity on the planet is increasing not decreasing [4] and the gap between rich and poor is growing [5]. Through to the early to mid 2000s the movement focused most heavily on individual action to address large "world problematiques" perhaps best typified by the work of Al Gore, with little international level success. This slow uptake, however, does not reflect lack of urgency or lack of will; rather, as we have argued elsewhere, scale is important in the area of environmental intervention [6] and the scale at which we engage with environmental issues is no different. International efforts can be powerful, as in the case of addressing the ozone crisis, but they are also slow and cumbersome and often either fail or become ineffectual. Individuals, on the other hand, have good control over certain elements of their lives, such as purchasing power, but have little to no direct control over urban planning and energy supply. Environmental action at the very large and very small scale might now be revealing its intrinsic limits. Indeed, the further we get away from an individual tending his or her own garden, the less effective planning and management decisions are, yet the probability of achieving sustainability decreases at finer scales [7].

More promising efforts are being seen at the community scale. In our research tracking positive community level efforts to encourage sustainable development in a wide variety of fields such as transport, energy, and infrastructure, we found examples that strongly suggest that it is at the community scale that the application of innovation, both technological and social occurs most effectively, and, when aggregated has the greatest impact in increasing sustainability at a broader scale. It is this scale therefore that is most important in the struggle to "craft" a more sustainable world. Communities can be defined broadly, not only by place, but also overlapping communities of practice [8], professional affiliation, shared interests and networks [9], and space, that is, virtual communities. In addition, the label "community" requires that the constituent population has formed a regularly interacting system of networks [10]. This research focuses on "communities of place" [11] as being where the interface between social capital and the environment occurs, but it also recognizes that virtual communities have great influence on the place based actions and innovations that result in sustainable community development.

Unfortunately communities in both rural and urban settings have been under unprecedented attack in the second half of the twentieth century [12,13]. Single industry and resource towns have been hit hard by the globalization of the economy and policies to create highly skilled and mobile workforces, while arguably increasing the economic opportunities for individuals, have worked against the stability and social health of communities. Planning orientated around car mobility rather than people has created infrastructure and places where chance social interactions are reduced, where people are isolated from the natural world and where streets and downtowns are increasingly empty places [14]. Such planning has also increased the homogeneity of residential areas, decreasing social-economic and cultural diversity in these places. All these developments make it harder for communities to thrive.

Some communities, however, remain strong in the face of external challenge. They possess what we call "community vitality"; they are resilient, they are innovative, and they are adaptive. Simply put, a vital community is one that can thrive in the face of change. It is a place that can remain at its core a functional community without loss to ecological, social and economic capitals in the long run, whatever occurs as a result of exogenous changes beyond its control. And perhaps more importantly, it is a place where human systems work with rather than against natural systems and processes.

## 4.2 LESSONS FROM SUSTAINABLE DEVELOPMENT

Community acts as a stage for environmental intervention and as a support network and empowering agent for those who wish to address environmental issues. Established communities have a sense of place; place has emerged as a feature of sustainable communities and sustainable development projects have been proposed to strengthen sense of place. Several writers have explored the importance of place, and a sense of place, growing from work pioneered by early writers in the area of human geography [15-17]. This dialogue has grown to include discussion of quality of life, the liveability, and the sustainability of human communities [14,18]. In contrast, the suburban form that arose in the latter half of the twentieth century embodies a placelessness summed up well by Debord's description of the suburban landscape as conforming to the motto "on this spot nothing will ever happen and nothing ever has" [19]. Orr maintains that the weakening sense of place is at the heart of our ecological crisis [20]. It is possible place is a necessary condition for the implementation of sustainable development.

The discourse of sustainable development itself has also changed since its broad scale recognition in 1987 through the publication of the Brundtland Commission Report. The early conceptualizations of sustainable development were very goal oriented and the movement now is much more process-oriented. As Holling argues, "sustainability is the capacity to create, test, and maintain adaptive capability. Development is the process of creating, testing, and maintaining opportunity. The phrase that combines the two, 'sustainable development' thus refers to the goal of fostering adaptive capabilities and creating opportunities. It is therefore not an oxymoron but a term that describes a logical partnership" [21].

Treating sustainable development as a process creates the need for an indefinite program of monitoring and adjustment. Every successful adaptation is only a temporary "solution" to changing selective conditions [22]. In short, sustainable development is a moving target. In some cases, the time spans involved are long to the point of being indefinite. This need for a continuous process arises due to two factors; the inherent unpredictability of complex adaptive systems, and the changes brought about by human innovation. This approach is a shift from a command and control model of sustainable development to a self-organizational model of dynamic sustainable development, a model more suited to the community scale. Such a model is more likely to be successful as it can emerge organically from unsustainable behaviour in manageable steps. Norms cannot be imposed in advance [23], but will emerge as part of an adaptation process. Instead of being a final objective, sustainable development has to be understood as a continuous process of change [24], and a fruitful approach to this process is to treat it as an evolution [22]. This shift to a concept of sustainable development as an evolving target explains why adaptability and innovation are as important as resilience, and why the early environmental movement's focus on looking back to a simpler time was not a successful strategy. Sustainable communities need the ability to embrace change and the tools to address such change.

Results from the previous five-year research agenda found that place [25] matters deeply to many Canadian communities, but there is little awareness of the aggregate impacts of human scale [6], the need for limits [26] and their subsequent impact on diversity [27], particularly biodiversity.

## 4.3 RESEARCH METHODOLOGY

This article builds upon the investigations of the Canada Research Chair (CRC) research program led by the first author from September 2004–September 2009, continuing to use a mixed-methods and contextual, comparative case study [28] approach. Case study methodology is particularly useful for addressing questions regarding the how and why of phenomena and in providing details about specific behaviours, a particularly necessary approach for the exploration of community vitality. The great strength of the case study is it provides a sense of context and a richness of detail that exceeds virtually every other approach to analysis [29]. Each case study, individually and in its contribution to larger analysis, acts as a heuristic for interrogating larger theories. A contextual comparative case study examines the commonalities and difference in the events, activities, and phenomenon that are the units of analysis in a typical case study. The purpose of engaging in cross-case or meta-case analysis is to enhance the researcher's capacity to understand how relationships may exist among discrete cases in order to refine concepts and build or test theory. Yin [28] adds that case study methodology is well suited to "how" questions. The approach of using multiple settings allows for the data source triangulation explained by Denzin [30] in which the research compares the data generated in different contexts. Yin [28] suggests two principles of data collection that were used for this research: the use of multiple sources of data and the creation of a case study database (see www.crcresearch.org which contains summaries of all the case studies referred to in this article).

The case studies were chosen based on a number of criteria. All except on Australian example they were selected as leading Canadian examples of the development of sustainable infrastructure and the process sustainable community development. The cases in sustainable infrastructure were selected based on three key attributes: integrated planning, transformation and innovation, and transferability, as well as for scalability, adaptability, and resilience. The overall set was selected for a diversity of geographical region, economy, and project and community sizes within the Canadian context. Cases studies in sustainable community development were selected as they demonstrated explicit (and implicit) links among and between four substantive "pillars" of the Canada Research Chair in Sustainable Community Development—place, scale, limits and diversity.

Each case study was developed using a variety of sources in order to triangulate the information. Data was drawn from other published information, internal documents and web-sites, and interviews. The precise nature of these varied from case to case depending on the specific context and nature of each case.

For the purposes of this discussion, a meta-analysis of the thirty-five case studies from the first five-year research program were analyzed to indentify characteristics of community vitality common to all the cases. The data was then analysed for emergent common themes. Critical success factors  from each case were identified and then categorized according to the emergent themes. The nascent characteristics derived from this meta-case study analysis are detailed below.

We hypothesize that a vital community practises some form of what we have identified as anticipatory governance. Humans group for social reasons, of course, but also group in order to tackle challenges that are beyond the scope of individual action. A community that has richer groupings for the first reason is surely better placed to respond should the second condition arise. Communities are both proactive and reactive; ideally they plot a course forward in order to achieve common goals, and at the same time they are ready for any challenges that come their ways. Living beings and complex systems are influenced by and adapt to their surroundings. Communities are no different. They are always influenced, and changed, by their surroundings. Sustainable communities adapt and work within their environment rather than against it. A community that is "vital", however, does more than adapt and mitigate, it anticipates, designs and redesigns as it adapts. Or, if it cannot or fails to anticipate, then it contains within it the diversity and redundancy necessary to adapt in a way that prevents harm. This balance of adaptation and resilience creates communities that are living complex adaptive systems, changing as needed yet maintaining their identity.

Previous research reveals that communities currently face an array of social, ecological, and economic challenges, and their response to these challenges is mixed; while some communities struggle to survive, others thrive [31]. Understanding community vitality, why some communities are resilient, adaptive and innovative in the face of change and others are not, is a pressing research question. We are assuming that at least some degree of community vitality is necessary to stimulate the creativity, partnerships and trans-disciplinary relationships we have established are necessary for sustainable community development. We also suspect there is a strong place consideration—perhaps the spaces in which social interaction can occur, perhaps the invisible influence of dominant ecological features on creativity and thought within a community. In effect, a lack of vitality is a form of poverty that will mitigate against the development of these aspects of sustainable development. If this is the case then there should be some evidence of vitality within each and every case study. That does not necessarily mean that every community represented by a case study could be described as vital, it is also possible that vitality could be created through the partnerships developed by the sustainability project itself.

## 4.4 COMMUNITY VITALITY AND RESILIENCE

What exactly is resilience? Walker describes resilience as "the capacity of a system to undergo change and still retain its basic function and structure" [32] (online), an ability that is partly manifest through the proper functioning of governments. One definition of resilience is "the ability of groups or communities to cope with external stresses and disturbances as a result of social, political and environmental change" [33]. Resilience has also been defined as the capacity to deal with complex issues widely dispersed across a set of loosely connected actors [34], a definition that speaks to the collective. However, resilience is also a function of the social networks contained in a community. Resilience emerges from intra-scale and cross-scale interaction, however understanding the nature of resilience across scales is difficult because of dominance of different processes at different scales, non-linearity, and emergent properties [35], as well as human dynamics. Social resilience can be measured by proxy, using indicators such as the variability of income, stability of livelihoods, wealth distribution, and demographic change [33] and agency [36].

One of the ways communities respond and exhibit resilience is through their ability to innovate. Innovation is more than new technology; technical ingenuity creates new technology, but social ingenuity reforms old institutions and social arrangements into new ones [37]. Innovation in a complex society occurs, however, on many scales. At the smaller scale we see incremental innovations, which are small refinements that occur relatively continuously. Such sudden shifts can provide new technologies to protect ecosystems, can shift our resource use from one resource base to another, and can also increase our impact on ecosystems in new and unexpected ways. Somehow we have to have some idea of what effect an innovation might have.

Incorporating innovation into a model of sustainable development is difficult. Though technology can be seen as an "adaptive answer" to problems [22]; there is inherent uncertainty in the predicted outcomes of innovation [38]. For example, expectations of the computer revolution were a significant reduction in the use of paper, when in fact the opposite has occurred, a significant increase in its use. Innovations can give rise to new needs, but they introduce variation and learning that is essential to the exploration and development of new possibilities [39]. Some of our problems require systems innovations which enable the fulfillment of needs in an entirely new manner, yet planning is difficult when things useful to us today may be of no use in the future and things we do not value may be essential to humans living in the future [40]. This is the connection between adaptation and innovation, the latter is a sufficient and necessary continuation for the former. Our ability to use innovation can be described as our ability to be adaptive.

Diversity is also key to all three anticipated heuristics of community vitality—resilience, innovation and adaptation. Thus, it is keystone to both resilience and a community's ability to adapt. With respect to innovation,

Hamel [41] argues that strategic innovation is the result of bringing a diverse set of voices into the strategy dialogue, among other issues. Further, there is evidence that minority opinions stimulate creativity and divergent thought which, through participation, manifest as innovation [42]. What, then, does the meta-analysis of the thirty-five case studies reveal about community vitality, resilience and innovation?

## 4.5 NASCENT CHARACTERISTICS

A qualitative meta-case analysis [28] of the thirty-five case study communities reveals common features we assume are characteristic of community vitality. A summary of the thirty-five case studies is provided in Table 1. Each case was read through and key elements of the case were extracted into a table and then categorized. These categories pertained to the characteristics of the case, and in particular to the elements identified as critical success factors. The categories that were identified in the majority of the cases, and which pertain to the community context in which the case was situated were those described below. Other categories which either didn't relate to the majority of the cases, or which are not relevant to the community context included technological innovation, focus on food, ecological conservation and protection.

**TABLE 1:** Case study summaries, the detailed cases can all be read at http://crcresearch. royalroads.ca/case-studies/case-studies.

| Case Study | Case Summary |
|---|---|
| A Microgeneration Strategy for Canada | This case provides an overview of the potential for microgeneration energy in Canada. It examines the opportunities that microgeneration represents, and argues that this opportunity is being taken by other jurisdictions, while Canada lags behind. |
| Deep Water Cooling | This case compares deep water cooling systems in Halifax, Nova Scotia and Toronto, Ontario, and describes their ecological and long term economic benefits. |
| Energy Efficiency for Homeowners | This research examines why homeowners took part in the EnerGuide for Houses program in Halifax, Nova Scotia, and what were the barriers to participation. |

**TABLE 1:** *Cont.*

| Case Study | Case Summary |
|---|---|
| Energy Performance Contracting | The City of Toronto, Ontario has actively provided support for the use of energy performance contracting (involving comprehensive energy and water retrofits and building renewal initiatives) with respect to both private and public buildings located within the City. |
| Renewable Energy on Prince Edward Island | Despite its population of just 138,000, the Province of Prince Edward Island has undertaken an ambitious renewable energy strategy that has delivered innovative policies, public engagement strategies and economic benefits. |
| Wind Power Generation | Several initiatives are proposed that directly link wind power to the needs of nearby communities, such as the Wolfe Island Wind Project at Kingston, Ontario. |
| EcoPerth | EcoPerth is a non-profit organization that was created in 1997, primarily to address climate change issues within the town of Perth, Ontario (population approximately 6,000) and the surrounding rural area. |
| Mid-term Objectives: An Urban Experience. Toronto, Ontario | The City of Toronto, Ontario in 1990 committed to reduce carbon dioxide emissions by 20% by 2005, relative to 1988 levels. To implement these mid-term objectives, the City has put in place several mechanisms including: The City of Toronto's Energy Efficiency Office (EEO); and the Toronto Atmospheric Fund. |
| Towards Green Buildings: Calgary | The City of Calgary, Alberta was the first jurisdiction in Canada to adopt a sustainable building policy in 2004, a policy that, amongst other things, commits all City-owned building developing new and under-taking major renovations of occupied facilities to meet or exceed the silver level of the Leadership in Energy and Environmental Design (LEED) standard. |
| United We Can | In five years, United We Can, a downtown eastside Vancouver, British Columbia recycling project, evolved from a loose ad hoc network of "binners" (dumpster divers) into a thriving business enterprise and an increasingly healthy community of workers engaged in providing an essential recycling service to their broader community. |
| Long Term Planning Initiatives | The case examines three cities with different approaches to long-term planning. Edmonton, Alberta has a fiscal approach, considering the costs associated with the replacement of current infrastructure and setting out strategies to manage the replacement over time. Ottawa, Ontario and Calgary, Alberta both start from a vision document for the city involving community participation and long-term planning horizons. Ottawa employs "Smart Growth" principles and Calgary uses the "Triple Bottom Line". |
| Triple Bottom Line in Practice: From Dockside to Dockside Green | This case explores the planning process that has led to the re-development of the Dockside area of the City of Victoria, British Columbia. The adoption of a tendering process for potential developers based on Triple Bottom Line (TBL) methodology has meant that smaller, more progressive development companies were able to compete for the land, although the social imperative was comprised in the long term. |

**TABLE 1:** *Cont.*

| Case Study | Case Summary |
|---|---|
| What Makes a City Liveable? | This case looks at two communities of very different sizes, the Town of Okotoks, Alberta and the City of Vancouver, British Columbia, both of which have been attempting to implement development based on quality of life and sustainable development for a number of years. |
| Alternative Road Allocations, Whitehorse | This case study examines the practice of converting existing four-lane roadways to multimodal two-lane roads, often referred to as alternative road allocations using Fourth Avenue and Quartz Avenue, in Whitehorse, Yukon Territory to illustrate the process. |
| Integrated Transportation Strategies | In 2002, the town of Mont Saint-Hilaire, Quebec put in motion the development of a multi-functional suburb focused around a new heavy-rail commuter station providing service to downtown Montreal, Quebec. |
| Mobility HUBs, Toronto, Ontario | The concept behind the New Mobility HUB project is to fill in these gaps with a network of hubs across Toronto, which link multiple modes of sustainable transportation. |
| Sustainable Transportation | The case study examines whether mass transit systems can be used as a tool to encourage the development of sustainable communities. The case examines a proposed expansion to the Montreal, Quebec commuter rail system. |
| Green Waste Programs | This study focuses on two examples of organic waste collection—one province wide in Nova Scotia and one city wide in Whitehorse, Yukon Territory. These were chosen to provide two examples contrasting provincial with town scale systems, and where the collection stands alone or is integrated into a comprehensive waste management strategy. |
| Storm Water Management | This case study documents some of the innovative approaches being undertaken to mitigate contaminated urban storm run-off in Chilliwack, British Columbia, and Toronto, Ontario. |
| Airshed Improvement: Stakeholder perspectives | The Quesnel, British Columbia Air Quality Roundtable is implementing a consensus based airshed management plan based on results of a comprehensive air quality assessment completed by the BC Ministry of Environment. |
| Banking Community Assets | Local models for community based economic development are starting to emerge. BCA, is one such initiative, a community venture finance group located on Cape Breton Island in Nova Scotia, Canada. It was established in 1989 in response to the community's need for economic development. |
| Carfree Markets | This study investigates a local sustainable development initiative to establish a pedestrian zone within a Canadian urban community, Kensington—a neighbourhood in Toronto, Ontario |
| Community Action on Salt Spring Island | This story concerns the efforts of the activist community on Salt Spring Island, British Columbia to protect their sense of place in response to a large land purchase and a subsequent program of extensive logging and land clearance in critical watersheds. |

**TABLE 1:** *Cont.*

| Case Study | Case Summary |
| --- | --- |
| Community Engagement in Whistler2020 | This case study examines the key elements of the Whistler2020 (a planning visioning document produced by the Resort Municipality of Whistler, British Columbia) engagement process and analyses the reflections of 14 community leaders representing various sectors on their involvement in the plan. |
| Farmers' Markets and Local Food Systems | There is a movement towards strengthening the local food system on Vancouver Island, British Columbia. This case study addresses a key component of the local food system: food distribution by local agricultural producers. In particular, it concentrates on farmers' markets, an important aspect of food distribution. |
| GHG Reduction Recommendations in the Personal Transportation Sector | This study proposed various short, medium and long-term recommendations on how to reduce greenhouse gas emissions in the personal transportation sector. The recommendations were based on information gathered through extensive literature research and interviews selected based on their expertise in the personal transportation field, and consisted of people from government, non-government organizations and educational institutions. |
| Green Urban Infrastructure Assessment | GUIA is a software tool that provides a process to identify green infrastructure for urban municipalities in Canada. |
| Maleny | The relationship between social capital and sustainable development is examined focusing on the nature of development in a small community. Maleny is a small town 90 km north of Brisbane, Australia. Formerly a dairy farming area, it underwent a major transformation with an influx of new residents in the 1970's. The study documents a clash between different notions of development in this particular community. |
| Merritt | This case study examines the relationship between how a community feels about the characteristics of place and social capital. Specifically, it considers the spatial aspects of the small community of Merritt, a rural town located in the Nicola Valley of southern British Columbia, Canada. |
| Quest Food Exchange | Quest Outreach Society is a Vancouver, British Columbia-based organization that intercepts, processes and then redistributes non-marketable food to social service agencies and others in need in the region. |
| Salmon River Watershed Management Plan | The Salmon River Watershed Management Plan partnership started out as a group of stakeholders with a desire to produce an effective management plan to protect and conserve one of the few remaining watersheds in the Greater Vancouver Regional District in British Columbia that is still able to support productive fish stocks. |
| Sustainable Community Planning: Comox Valley | This case study explores issues related to planning for rapid population growth and implementing sustainability in community planning for the Comox Valley, British Columbia. |
| The National Round Table on the Environment and the Economy (NRTEE) | This case study examines the creation of the National Roundtable on the Environment and the Economy, Canada's first national multi-stakeholder process, the challenges it faced and its evolution over time. |

**TABLE 1:** *Cont.*

| Case Study | Case Summary |
| --- | --- |
| Trust for Sustainable Forestry: Cortes Island | This case study describes the creation and the first project of the Trust for Sustainable Forestry, a small not for profit trust created to develop small ecologically sensitive small communities in protected, but working forest environments. |
| Urban Food Distribution Systems | 13 examples of direct marketing methods in the delivery of farm to consumer food distribution. |

*Community Openness and Trust:* Trans-disciplinary partnerships and alliances are a very common aspect of sustainability projects. All the thirty-five case studies involved formal trans-disciplinary partnerships, normally involving the public and either the private sector, and/or civil society groups as well. Private/public partnerships are commonly referred to as P3 or PPP partnerships by governments; see for example Infrastructure Canada's website [43] or the UK's HM Treasury [44]. In terms of vitality this shows that where there is openness and communication flows rather than hostility between sectors the community is one that foster innovation and creativity. This is demonstrated, for example, in the case of Deep Water Cooling in Toronto, where a public private partnership enabled the co-operation and investment required for the project—the cooperation of multiple private sector organizations with the City provided the necessary economies of scale that made the cooling infrastructure a sensible investment, and the energy and cost savings sufficiently short to make economic sense. In the case of the EnerGuide for Houses in Nova Scotia the loss of federal support proved a significant problem to the long-term maintenance of the project, with the lack of federal grants to householders to support energy efficiency retrofitting meaning householders lacked incentive to examine their home's energy efficiency and possible retrofitting opportunities.

In the case of the Salmon River Watershed management plan failure to produce a robust management plan was, in part, attributed to the souring of relationships between stakeholders due to perceptions of vested interests and hidden agendas—directly impacting trust between parties. Thirty-two of the case studies exhibited facets of trust (or the lack thereof) that directly impacted the full realization of sustainable development in these various contexts.

*Connection with People and Place:* through a sense of the meaning of the place within the community, for example, the case of Salt Spring Island, Vancouver, British Columbia stimulates community attitudes and values to development that keeps the ecology as the basis of community action—either explicitly on Salt Spring Island in that a threat to the watersheds initiated very strong community response, or as in Okotoks, Alberta, where a proactive response to the possibility of future over-consumption of a natural resource (access to water) stimulated action. This could be seen as both utilitarian and duty-based ethical philosophies that both serve to initiate place-based community action. Planning initiatives in the Vancouver case study were almost entirely instigated by a desire to preserve access to key landscape features and the city, which has contributed to the ecological and social vitality of this city, now billed as one of the most liveable cities on the planet. Where the relationship to place is completely urban (for example in the Kensington Market case study in urban Toronto or Downtown Eastside in urban Vancouver, case studies), the connection to place tends to be not with the built environment, but to the people and social capital in the specific locale, and is manifest by the generation of networks of empowerment. In the United We Can case study it is connection through the empowerment of marginalized individuals through new network formation, and in Kensington, the creation of community identity with the market. All the case studies involved some connection to either community (69%), or natural place (51%), with fourteen case studies (40%) having both characteristics.

*Continuity and Stability:* Both stable leadership and stable funding are important in the case studies. In the EcoPerth case study, the leadership of the project was consistent and stable—at the same time the initiative engaged a diversity of people from the community, and was open to many influences. In the case of Kensington market, continuity of funding was identified as key in protecting the leadership from the constant stress of fund raising and therefore from burnout—this in turn allowed for stability of leadership as the core group was maintained, often solely lacking in civil society organizations, especially grass-roots and smaller groups. This stability in many ways contributed to the freedom that these projects had to engage with a greater diversity within (bonding social capital) and in some cases, outside the community (linking social capital), thus enhancing community vitality which built over the life of the project expanding the response from a climate change project to a broader program of sustainable development initiatives at the local level. This maintained community vitality in contrast to the experience of the Dockside Green case study in Victoria, British Columbia, where the municipal leadership individual within the city planning department that first supported the project left, and the knowledge lost meant that the project had reduced opportunities for integrating the ecological and particularly, the social imperatives. The loss of funding in the Nova Scotia EnerGuide program also affected the stability of the project and therefore its contribution to sustainable development when Federal funding was cancelled.

A balance between continuity and openness, therefore, appears to be an essential link to vitality. Security and stability of leadership, and partnerships, particularly private/public, enlarges the public sphere to pursue innovation and creativity. We believe both of these variables are directly linked to community vitality.

A lack of stability of population can also lead to barriers to community vitality. Cities with rapid growth or rapid turnover of population often struggle to create the stability that stimulates vitality. This is very apparent in the case of Wood Buffalo (Fort McMurray), Alberta where economic and social change has been so great that it has inhibited adequate forward planning, creating social and economic instability. This is not necessarily a function of the tar sands extraction per se, but the inability of a rapidly growing (unstable) population to create a shared vision of community. A less extreme example of this rapid population growth causing problems at the community scale can also be seen in the analysis of community planning in the Comox Valley of British Columbia where the fluid and frequently changing composition of municipal councils led to a loss of vision and inability to put in place robust and long-term planning policy.

*Perturbation:* Many of the case studies commenced or were instigated after a period of change, or perturbation to the status quo—the perturbation in these cases stimulated the innovation and creativity leading to the community action, notable examples from the case studies include the case examining community action on Salt Spring Island, where the change in management of critical ecosystems on the island stimulated the community response. Direct Marketing of food is a response to the decline of small farming and increasing barriers for small farmers to access markets. Also many of the cases, particularly the renewable energy and municipal planning related ones were responses to change—either locally through population increase for example as in Comox and Okotoks, or in a wider context with local responses to global climate change as in the case of Perth, Ontario and the various alternative energy projects examined.. It should be noted, however, that in many situations of perturbation, especially single-resource economy communities, where the economy changes, communities collapse—this is a phenomenon that the case studies did not examine. The research team will have to re-examine its criteria for case study selection to determine if a bias existed for success, and many research questions flow from this meta-analysis finding, outside the scope of this paper.

It is indeed plausible that given the work of C.S. Holling and the Resilience Alliance (http://www.resalliance.org/) that perturbation is necessary for the maintenance of vitality, with the absence of perturbation leading to stagnation. As the Holling model suggests, moving from exploitation to conservation to renewal to release is the structure of ecosystem functioning. Dale, however, has noted that in human activity systems, especially governments, the pattern is "stuck" in oscillating between exploitation and conservation, with very little release and no renewal [45]. Again, the meta-case analysis has revealed a discrepancy, in that all the case studies which examined some aspect of governance, planning, or the adoption of technological solutions, show innovative and creative re-organization of some type. This reorganization is apparent at the Federal level with the case of the NRTEE, Canada's first national multi-stakeholder process, in response to the Brundtland Commission report. At the more local or regional level, many of the cases involved the creation of new trans-disciplinary stakeholder groups, for example in Quesnel, Salmon River, Merritt and Whistler. Adoption of new bylaws or policy was also evident to facilitate sustainable development—particularly in the creation of new zoning types on Cortes Island to allow "ecovillage" style development or the creation of limits to growth in communities such as Okotoks and Whistler to protect the natural resource base. The antecedents need to be further explored.

This finding, that is, perturbation, apparently contradicts the outcome that stability is also important and that too much change, as in Fort Mc-Murray, inhibits vitality. Similarly, there may be a link between the degree in which a community can respond to change and its functional social diversity—assuming that social diversity is analogous to functional ecological diversity defined as "the range and value of those species and organismal traits that influence ecosystem functioning" [46]. Broadly speaking, the more complex and diverse a system is in terms of the functional groups it contains, then the greater degree of functionality it manifests—greater functional diversity leads to greater stability of the ecosystem, although this may be at the expense of the stability of the abundance of individual species within the ecosystem.

If the perturbation, however, is happening in a way that maintains core stability then it actually stimulates vitality. For example, in Maleny, Australia, the threat of development on valued open land in the community led to a collective vision in a previously divided community between outsiders and long-term residents, with the resultant campaign bringing newcomers and the core community together, an example of the creation of greater community vitality as a result of a perturbation. It may be that community vitality is related to degree of community cohesion, and there may also be an integral relationship between adaptive governance, stability and community vitality that will be explored further in the next five-year research program.

The community response in Salt Spring Island, Canada, arose from a major threat to key watersheds and a change in land ownership, mobilizing the community existing social capital, stimulating the development of greater bridging social capital between disparate networks. Similarly, the BCA program in Nova Scotia was a response to economic under development and the threat of outmigration, which led to the development of enhanced community vitality through economic diversification through co-operative ventures within the community.

It seems clear that a perturbation is needed to stimulate action, and in some cases, vitality, which may be the explanation for why many municipal governments only react to change and do not very often, predict and anticipate change [47], a too stable status quo decreases vitality. It appears as if communities need the change, or the exogenous shock, to "loosen" innovation and creativity, which in turn stimulates vitality. This is analogous to Holling's creative destruction or release leading to renewal. The changes are necessary, and the clue is to build redundancy at the local level and resilience to buffer especially exogenous shocks, so the change is not catastrophic. This supports the importance of both variables as necessary for sustainable community development as the change contributes to enhanced vitality, assuming vitality and sustainable development themselves are linked.

Although 42% of the case studies can be directly attributable to stimulation from a perturbation, they are disproportionately grass roots or small community case studies. It seems that internal (community or institutional) capacity and, likely diversity of networks and resources, increases the capacity of creative and innovative action and thought. Again, referring to Holling's work, rigidity of institutional and community responses may be more likely in larger scale than smaller communities, similarly, the capacity of what we have defined as adaptive governance in the next five-year research program. Perhaps, this increased capacity to respond, as a function of size, increases the ability to more quickly perceive larger and wider scale perturbations and how this may impact or contribute to greater community diversification. An interesting question to explore will be to determine whether projects instigated by institutions contribute to community vitality in the same ways as grassroots projects, and to what extent scale (the relative proportion between community population and people involved in the project) and connectivity (the density and centrality of networks between the wider community and those involved in the project) contributes to enhance community vitality.

*Diversity:* All of the above suggests that diversity is also the (or at least one) of the basic components of community vitality, as it is for sustainable development [27]. Community openness enables and facilitates the transdisciplinary co-operation needed to implement sustainable development solutions, and the incorporation into the dialogue around such projects ensures a variety  and complexity only achieved through the innovation of socially diverse groups. The Salt Spring Island campaign particularly illustrates this, with the involvement of community activists, provincial organizations, the private, public and community sectors, rich well connected benefactors and low wages frontline protestors. However, some degree of more than normal diversity of interactions seems to be apparent in all the case studies. Broader and denser degrees of human-human connection as well as human-ecology connection increase the diversity of relationship within a community, and the broadening of the concept of community to include ecological relationships, necessary for sustainable development, but also perhaps for vitality. This may become increasingly important if, as commentators such as Kunstler and Rubin predict, peak oil means economies become more localized.

## 4.6 CONCLUSIONS

The most common characteristic of the these case studies that represent the first phase of our research is that all of the thirty-five case studies demonstrate evidence of partnership of one form or another. Given the complexity of implementing sustainable development, its cross-sectoral, interdisciplinary aspects and its cross-jurisdictional institutional focus, this is perhaps to be expected. Moreover, resilience theory suggests that key system components, and the focal scales at which they interact, are often best identified through strategies that partner experts with stakeholders who understand the system from different scales and perspectives [48]. Community vitality in the form of the willingness and agency to form partnerships is a key element of successful sustainable development.

In addition, there appear to be key relationships between partnerships and the ability to innovate. Since partnerships and strategic alliance can reduce the risks of innovation and the uncertainty surrounding the early take-up of new technologies, it would appear to be a strategic advantage to such relationships. As well, social capital and network formation appear in many of the case studies as a key characteristic which is also linked to the diffusion of innovation, since most people decide to adopt an innovation "primarily on subjective values and social norms diffused through interpersonal networks, rather than as a result of rational reflection on scientific data" [49-52].

Our work over the last five years has demonstrated that the community scale acts as an important locus of sustainable development diffusion. Community vitality both provides the needed resilience to weather social, economic, and environmental change, and also provides a site for innovation where problems can be addressed iteratively with a process-based approach through the active engagement of diverse social actors. Community vitality, however, has been badly damaged in the industrial world by the suburbanization of the second half of the twentieth century. Trans-disciplinary dialogue that builds on the need for openness and in-creased trust within communities, both place based and virtual, may assist in revitalizing community, but our research shows barriers to the collective solving of difficult issues on-line still persist [53]. Increasing community vitality may prove to be a strategic policy direction for governments in the process of sustainable development and a natural bridge between in-dividual action and action at the international and national scale. Further research is required to concretize community vitality as distinct from re-silience and sustainable development.

## REFERENCES

1.    Carson, R. Silent Spring; Houghton Mifflin Company: Boston, MA, USA, 1962.
2.    Brundtland, G. Our Common Future: World Commission on Environment and De-velopment; Oxford University Press: New York, NY, USA, 1987.
3.    Agenda 21: Earth Summit—The United Nations Programme of Action from Rio; United Nations: New York, NY, USA, 1993.
4.    Living Planet Report 2008; Hails, C., Ed.; World Wildlife Fund (WWF) Interna-tional: Geneva, Switzerland, 2008.
5.    World of Work Report 2008: Income Inequalities in the Age of Financial Globaliza-tion; International Labour Office: Geneva, Switzerland, 2008.
6.    Newman, L.; Dale, A. Large footprints in a small world: toward a macroeconomics of scale. Sustain. Sci. Pract. Policy 2009, 5, 1-11.
7.    Forman, R. Land Mosaics: The Ecology of Landscapes and Regions; Cambridge University Press: Cambridge, UK, 1995.
8.    Lesser, E.; Prusak, L. Communities of practice, social capital, and organizational knowledge. In The Knowledge Management Yearbook 2000–2001; Cortada, J.W., Woods, J.A., Eds.; Elsevier: Amsterdam, The Netherlands, 2000; pp. 251-259.
9.    MacKinnon, M.P.; Maxwell, J.; Rosell, S.; Saxena, N. Citizens' Dialogue on Cana-da's Future: A 21st Century Social Contract; Canadian Policy Research Networks: Ottawa, Canada, 2003.
10.    Onyx, J.; Osburn, L.; Bullen P. Response to the environment: social capital and sus-tainability. Aust. J. Environ. Manage.2004, 11, 212-219.
11.    Flora, J.L. Social capital and communities of place. Rural Sociol. 1998, 63, 481-506.

12. Jacobs, J. The Death and Life of American Cities; Vintage Books: New York, NY, USA, 1961.
13. Kunstler, J.H. The Geography of Nowhere: The Rise and Decline of America's Man-Made Landscape; Simon & Schuster: New York, NY, USA, 1993.
14. Hanna, K.; Dale, A.; Ling, C. Social capital and quality of place: reflections on growth and change in a small town. Local Environ. 2009, 14, 33-46.
15. Relph, E. Place and Placelessness; Pion Limited: London, UK, 1976.
16. Tuan, Y.F. Space and Place; Arnold: London, UK, 1977.
17. Seamon, D. A Geography of the Lifeworld; Croom Helm: London, UK, 1979.
18. Waterton, E. Whose sense of place? Reconciling archaeological perspectives with community values: cultural landscapes in England. Int. J. Herit. Stud. 2005, 11, 309-325.
19. Debord, G. Society of the Spectacle; Black & Red Publishing: Detroit, MI, USA, 1983.
20. Orr, D. Lessons from the edge. Alternatives 2007, 35, 40-52.
21. Holling, C.S. Understanding the complexity of economic, ecological, and social systems. Ecosystems 2001, 4, 390-405.
22. Rammel, C.; van den Bergh, J. Evolutionary policies for sustainable development: adaptive flexibility and risk minimizing. Ecol. Econ. 2003, 47, 121-133.
23. Robinson, J. Squaring the circle? Some thoughts on the idea of sustainable development. Ecol. Econ. 2004, 48, 369-384.
24. Jokinen, P.; Malaska, P.; Kaivo-Oja, J. The environment in an information society: a transition stage towards more sustainable development. Futures 1998, 30, 485-498.
25. Dale, A.; Ling, C.; Newman, L. Does place matter? Sustainable community development in three Canadian communities. Ethics Place Environ. 2008, doi:10.1080/13668790802559676.
26. Newman, L.; Dale, A. Limits to growth rates in an ethereal economy. Futures 2008, 40, 261-267.
27. Dale, A. Diversity: why is the human species so bad at difference? J. Urban Plan. submitted.
28. Yin, R. Case Study Research: Designs and Methods, 3rd ed.; Sage Publications: Newbury Park, CA, USA, 2003.
29. Merriam, S. Qualitative Research and Case Study Applications in Education; Jossey-Bass Publishers: San Francisco, CA, USA, 1988.
30. Denzin, N. The Research Act; Prentice Hall: Upper Saddle River, NJ, USA, 1984.
31. Dale, A.; Onyx, J. A Dynamic Balance: Social Capital and Sustainable Development; UBC Press: Vancouver, Canada, 2005.
32. Walker, B. Resilience thinking. People Place, 24 November 2008.
33. Adger, N. Social and ecological resilience: are they related? Prog. Hum. Geogr. 2000, 24, 347-364.
34. Olsson, P.; Folke, C.; Berkes, F. Adaptive co-management for building resilience in social-ecological systems. Environ. Manage. 2004, 34, 75-90.
35. Peterson, G. Political ecology and ecological resilience: an integration of human and ecological dynamics. Ecol. Econ. 2000, 35, 323-336.
36. Newman, L.; Dale, A. The role of agency in sustainable local community development. Local Environ. 2005, 10, 477-486.

37. Homer-Dixon, T. The Ingenuity Gap; Alfred A. Knopf: New York, NY, USA, 2000.

38. Buenstorf, G. Self-organization and sustainability: energetics of evolution and implication for ecological economics. Ecol. Econ. 2000, 33, 119-134.

39. Vollenbroek, F. Sustainable development and the challenge of innovation. J. Clean Prod. 2002, 10, 215-223.

40. Gowdy, J. The social context of natural capital: the social limits to sustainable development. Int. J. Soc. Econ. 1994, 21, 43-55.

41. Hamel, G. The challenge today: changing the rules of the game. Bus. Strategy Rev. 1998, 9, 19-26.

42. de Dreu, C.; West, M. In defense of the individual: the CEO as board chairperson. J. Appl. Psychol. 2001, 86, 1191-1201.

43. Frequently Asked Questions; Infrastructure Canada: Ottawa, Canada, 2009; Available online: http://www.buildingcanada-chantierscanada.gc.ca/resources/faq/faq-eng.html (accessed on 14 December 2009).

44. HM Treasury. Public Private Partnerships; Available online: http://www.hm-treasury.gov.uk/ppp_index.htm (accessed on 14 December 2009).

45. Dale, A. At the Edge: Sustainable Development for the 21st Century; UBC Press: Vancouver, Canada, 2001.

46. Tilman, D. Functional diversity. In Encyclopedia of Biodiversity; Levin, S.A., Ed.; Academic Press: Durham, NC, USA, 2001; pp. 109-120.

47. Dale, A. Governance for sustainable development: as if it mattered? In Innovation, Science and Environment 2009–2010. Special Edition—Charting Sustainable Development in Canada 1987–2007; Toner, G., Meadowcroft, J., Eds.; McGill-Queen's University Press: Montreal, Canada, 2008.

48. Westley, F.; Carpenter, S.; Brock, W.; Holling, C.; Gunderson, L. Why systems of people and nature are not just social and ecological systems. In Panarchy: Understanding Transformation in Human and Natural Systems; Holling, C.S., Gunderson, L.H., Eds.; Island Press: Washington, DC, USA, 2002.

49. Atwell, R.C.; Schulte, L.A.; Westphal, L.M. Linking resilience theory and diffusion of innovations theory to understand the potential for perennials in the U.S. corn belt. Ecol. Soc. 2008, 14, 30:1-30:17.

50. Ryan, B.; Gross, N. The diffusion of hybrid seed corn in two Iowa communities. Rural Sociol. 1943, 8, 15-24.

51. Coleman, J.; Katz, E.; Menzel, H. The diffusion of an innovation among physicians. Sociometry 1957, 20, 253-270.

52. Rogers, E. Diffusion of Innovations, 4th ed.; The Free Press: New York, NY, USA, 1962.

53. Dale, A.; Ling, C.; Newman, L. Facilitating trans-disciplinary research teams trough on-line collaboration. Int. J. Sustain. High Educ. In press.

## CHAPTER 5

# Cross-Scale and Cross-Level Dynamics: Governance and Capacity for Resilience in a Social-Ecological System in Taiwan

HSING-SHENG TAI

## 5.1 INTRODUCTION

Since the Brundtland Report [1], sustainable development and sustainability have been promoted worldwide as guiding principles for Earth's future. Numerous indicators, models, actions and institutions have been developed to make these concepts operational for the pursuit of sustainability goals. Although these efforts certainly constitute important starting points, questions are still raised concerning their fundamental limitations. Anderies et al. [2] argue "it is insufficient, and even dangerous, to assume that individual actions will aggregate up to generate system-level sustainability." The complex and uncertain nature of feedback systems demonstrates a need to understand the cross-scale and cross-level dynamics of the system as a whole. Otherwise, individual actions, however well-intentioned, are eventually futile.

Resilience thinking represents an academic paradigm shift that can help clarify the meaning of sustainability [3]. Resilience addresses the dynamics of complex social-ecological systems from a systems perspective [4,5,6]. By recognizing characteristics of complexity, uncertainty, nonlinearity, thresholds, irreversibility, and multi-scale and multi-level interactions in a changing world, resilience thinking goes beyond traditional resource management to embrace the dynamic and complex nature of social-ecological systems [7].

Rather than focusing on narrowly defined policy goals and seeking optimal states, resilience thinking stresses that human societies should manage general system properties with resilience, adaptability, and transformability in mind [4,5,6]. Resilience represents the capacity of a social-ecological system, when facing a new context, to continually adapt, maintain fundamental functions and structures, and remain within critical thresholds. Resilience, adaptability and transformability possess the nature of both conservation and creation. They operate and interact across multiple scales, giving insight into sustainability. Adaptability represents the capacity of social-ecological systems to adapt and to follow the current development trajectories while maintaining original system properties. Transformability is the capacity to cross thresholds into new development trajectories and new system properties when existing conditions are untenable. Human societies should foster the resilience, adaptability, and transformability properties of social-ecological systems to enhance Earth's system resilience [4,6].

To clarify confusion regarding sustainability and resilience, Anderies et al. [2] said sustainability "should be used to refer to an analytical framework to guide actions across all levels of organizations related to the way human societies operate and interact with their environments." This analytical framework includes two main components: sustainability measures of system performance and sustainability decision-making context. It is in the decision-making context, Anderies et al. [2] argued, that resilience thinking becomes critical and useful for adequately understanding processes associated with pursuing sustainability goals. Knowledge about the dynamics of social-ecological systems is central for developing governance structures that effectively contribute to achieving sustainability goals.

By applying a resilience-based analytical framework, this article focuses on governance issues by examining associations among cross-scale and cross-level dynamics, governance, and capacity to enhance resilience of a regional social-ecological system. Specifically, it addresses the following pivotal question: how do processes and patterns of cross-scale and cross-level interactions influence attributes of a governance system and strengthen or weaken its capacity for resilience? Therefore, I investigated the long-term dynamics of a social-ecological system: The Danungdafu Forestation Area (DFA) in Hualien, Taiwan. The DFA represents one of Taiwan's most controversial cases of land use, indigenous rights, and environmental issues [8].

This paper is organized into six sections. Section 2 reviews existing analytical frameworks on governance, cross-scale dynamics, and capacity for managing resilience. In this section, I recommend amendments to the existing framework. Section 3 describes the research approach. Section 4 presents historical, political, social, economic, and ecological settings from which the DFA case evolved. This section also examines the cross-scale and cross-level dynamics of DFA. In Section 5, I analyze governance system attributes and capacity for managing resilience, based on the modified framework in Section 2. In Section 6, I present the main conclusions and policy implications.

## 5.2 GOVERNANCE, CROSS-SCALE AND CROSS-LEVEL DYNAMICS, AND CAPACITY

Governance is a central issue of sustainability [9]. Any discussion of resilience must examine the capacity of governance to adapt to changing conditions [5]. Governance refers to the structures and processes through which human societies share power; shape incentives, behaviors, identity, and decision-making; interact with each other; and influence outcomes. Governance involves: (1) stakeholders and actors, including government, communities, businesses, and non-governmental organizations; (2) different institutions, both formal and informal; and (3) various actions and decision-making processes, including governmental and jurisdictional processes, debates, negotiation, mediation, public consultations, conflict resolution, elections, and protests [9,10,11].

In complex, multi-scale and multi-level circumstances, cross-scale and cross-level dynamics are recognized as core issues for governance of social-ecological systems [12,13,14,15]. Gibson et al. [16] (p. 218) define scale as "the spatial, temporal, quantitative, or analytical dimensions used to measure and study any phenomenon". They define levels as "units of analysis that are located at the same position on a scale". A number of scholars [12,13,14,15,17,18] have urged the need for scale- and level-sensitive governance approaches, calling for more in-depth understanding of the complex processes associated with scale issues. Cash et al. [12] concluded that people consciously dealing with scale issues are generally more able to successfully identify problems and seek sustainable solutions. Armitage [15] argued that, in a multi-level world, multi-level governance can contribute to adaptation capacity which is central to resilience. Young [14] focused on differentiating governance regimes among levels on the jurisdictional scale, identifying commonly interacting patterns, including dominance, separation, merger, negotiated agreement, and system change. He stressed that these interacting patterns often profoundly influence outcomes of governance in terms of sustainability.

Through a comprehensive review, Termeer et al. [17] compared three representative governance regimes from the perspective of scale: monocentric governance, multilevel governance, and adaptive governance. Monocentric governance exists when central state authorities are in charge of governing social-ecological systems. It mainly addresses jurisdictional scale issues. Multilevel governance extends its focus to multiple level interactions, specifically at jurisdictional and spatial scales. In this model, governance tasks allocated to different levels, whether involving public or private actors, need to coordinate with each other. Adaptive governance considers a whole spectrum of cross-level and cross-scale interactions. These interactions may involve jurisdictional and spatial scales, as well as temporal, institutional, management, network, and knowledge scales.

In the associations between governance and the capacity to manage resilience, capacity stands for the central concept of social-ecological system resilience. Capacity, as an umbrella concept, has been divided into detailed sub-concepts. Although extensively discussed in specific fields, academia needs a systematic framework for deliberate differentiation among, and exploration into, different dimensions of capacity that simultaneously addresses capacity's links to governance. Lebel et al. [11] may fill this gap. They propose an analytical framework for the ways governance attributes affect society's capacity to manage resilience in regional social-ecological systems. They identify three sets of governance attributes: (1) participatory and deliberative; (2) polycentric and multi-layered; and (3) accountable and just. They divide capacity into six different dimensions ([11], Figure 1): (1) scale, or the capacity "to engage effectively with and handle multiple- and cross-scale dynamics"; (2) uncertainties, or the capacity "to anticipate and cope with uncertainties and surprises"; (3) fit, or the capacity "to design institutions which fit diverse social and ecological contexts"; (4) thresholds, or the capacity "to detect and navigate hard-to-reverse thresholds"; (5) knowledge, or the capacity "to combine and integrate different forms of knowledge"; and (6) diversity, or the capacity "to maintain ecological and social diversity". Lebel et al. [11] found strong associations between these three governance attributes and the six capacities for managing resilience.

In an extensive review of the literature, Armitage [15] also summarized key attributes of adaptive, multi-level governance. Although Armitage did not differentiate capacity from governance attributes, his governance attributes may complement the framework proposed by Lebel et al. [11]. Specifically, Armitage's list of governance attributes includes participation, collaboration, deliberation, multiple layers, accountability, interaction, knowledge pluralism, and learning. These are attributes or capacities considered by Lebel et al. [11], although some differences in terms exist. On the other hand, three attributes summarized by Armitage [15], networked, leadership, and trust, are not explicitly addressed by Lebel et al. [11]. Given the importance of these three widely recognized features, I recommend amendments to the framework of Lebel et al. [11] by adding one more governance attribute, networked, and two more capacity dimensions: leadership and social capital. I use the term social capital instead of trust, because a social system needs trust to bond and bridge social networks. These networks and trust often constitute the concept of social capital [19].

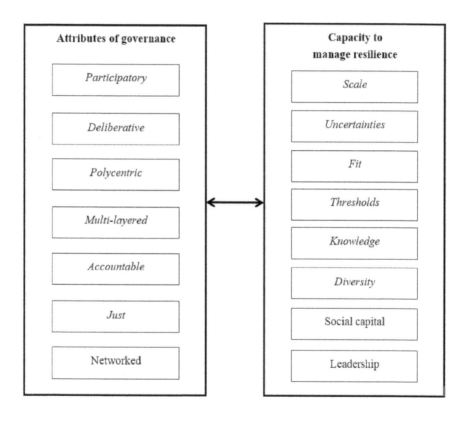

**FIGURE 1:** Framework representing associations between essential governance attributes and the capacity to manage resilience. Attributes and capacities analyzed by Lebel et al. [11] are in italics.

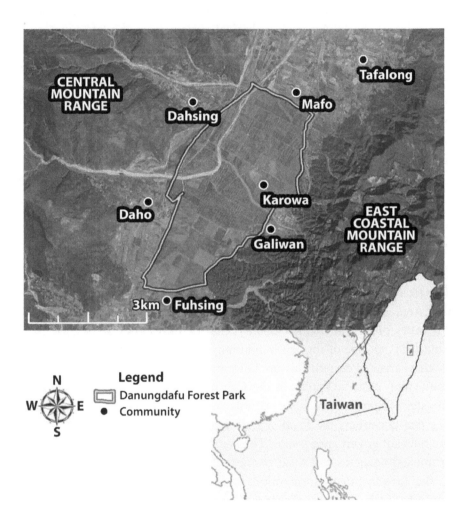

**FIGURE 2:** Map of the Danungdafu Forestation Area in eastern Taiwan. Satellite map data: Google earth 2013, 2013 Cncs/Spot Image Image 2013 DigitalGlobe.

## 5.3 RESEARCH APPROACH

This article traces and describes a long-term cross-level and cross-scale dynamics of the DFA social-ecological system from the late nineteenth century to present. For historical events, I drew on academic literature and official documents. The oral history method was adopted to include perspectives of marginalized social groups. To explore dynamics after the millennium, I used a variety of qualitative research methods, including in-depth interviews, focus groups, participant observations, and literature analysis. Field study spanned June 2012–June 2014. After summarizing the dynamic evolution of this system, I followed the modified framework shown in Figure 1 to describe attributes of the existing governance regime and its capacities for resilience.

## 5.4 CROSS-SCALE AND CROSS-LEVEL DYNAMICS OF THE DFA

The focal social-ecological system of this study, the Danungdafu Foresta-tion Area, geographically includes forest of 1138 hectares located in the rift valley of central Hualien County, eastern Taiwan (23°36′N, 121°24′E, Figure 2). As one of the few remaining forests in intensively developed lowland areas of eastern Taiwan, DFA may be a critical ecological corridor connecting protected areas in the Central Mountain Range and the East Coastal Mountain Range. As regards social system, the key characteris-tics that constructs the focal system is land rights debate, together with the relevant governance issues. The social system mainly consists of gov-ernmental agencies and local communities. Land property rights belong to the Taiwan Sugar Corporation, which has the central gonvernmental control over the majority of shares. The Forestry Bureau of central govern-ment takes charge of the management of forests. Buffering the DFA, there are seven local communities which are historically, socio-economically, culturally closely related to this piece of land. Populations of these com-munities has two main demographic sources: indigenous Amis people and Chinese-descent Han people. Some indigenous communities near DFA advocate that DFA is a traditional territory of social, cultural and economic significance. Therefore, they accordingly claim resumption of the land property rights [8]. Han people generally do not claim land property rights, but have a high expectation of being able to benefit from utilization of land and forests. Although the two populations have different views concerning land rights and different origins of cultural identity, they do share similar sense of belonging to the land of DFA, and are highly interested in governance issues. As Subsection 4.1, Subsection 4.2, Subsection 4.3 and Subsection 4.4 will show, such a sense of belonging is deeply rooted in the population's long-term connections to the land. Even though the ecological facet of the system had experienced two major transformations in the past century, the governance structure does show high continuity, in which local communities struggle to improve their well-being, and complex interrelationships among various communities and governmental agencies continue to exist. It is in this context that the land and forests of DFA, local communities, and relevant governmental agencies constitute the focal social-ecological system that is under investigation.

Given the controversial nature and importance of DFA, the following question continues to be the focus of debate: how can DFA be governed to achieve sustainability goals? To answer this question, the first step is to identify main governance issues, drawing on full understanding of context, system dynamics and status quo [20]. Cash et al. [12] illustrated some scales and levels critical for understanding social-ecological system dynamics. Of these, spatial, jurisdictional, institutional, networks, and knowledge scales are closely related to the case of DFA. Relevant scales and levels are listed in the Table 1. I analyzed the dynamics of DFA following these subsections.

**TABLE 1:** Scales and levels critical for understanding the Danungdafu Forestation Area (DFA) case.

| Scale | Levels (from low to high) |
|---|---|
| Spatial | Patch, region, nation, globe |
| Jurisdictional | Community, regional, national |
| Institutional | Local institution, laws and regulations, constitution |
| Networks | Kin, community, bridging organizations |
| Knowledge | Specific, general |

## 5.4.1 FROM FORESTS TO SUGAR PLANTATION; 1895–1945

Prior to Japanese rule of Taiwan from 1895 to 1945, the DFA was covered by bush forests interspersed with rivers and ponds. It provided wildlife habitat for large mammals, such as deer and wild boar, that were frequently hunted. Indigenous people of the Amis tribe, particularly the Karowa community, lived there, engaged in slash-and-burn farming, and hunted to sustain livelihoods and maintain their traditional common-pool resources regime [21,22]. Another Amis community, Tafalong, also used this land for livelihood and cultural festivals.

The shift in 1895 to Japanese rule fundamentally changed both human communities and the landscape of the DFA. The Japanese government regarded Taiwan as a source of natural resources and commodities production base for both domestic consumption and international trade. Cane sugar, a lucrative commodity at that time, required large areas of land. This constituted the main motive for legislation at the early stages of Japanese rule. New laws nationalized property rights of all "ownerless" land in Taiwan, most of which was Taiwan's indigenous people's traditional territory [23]. After nationalization, the government leased, under favorable terms, large areas of land to private enterprises, encouraging them to develop the land. This is the context of eastern Taiwan's inclusion in the development of a modern state [24].

In 1921, a Japanese private enterprise, Salt and Sugar Ltd., acquired use rights of DFA and began to expel the Karowa community of Amis people from DFA. Although indigenous people resisted the encroachment of these outsiders, their actions were futile under the massive pressure of the public authority of the government [22]. People in the Karowa community were forced to relocate to marginal lands and other indigenous communities around DFA, as well as other places in central Hualien County. The community virtually collapsed and lost most of their social networks and identity. Poor conditions of marginal lands contributed to the deterioration of livelihoods. The effects are still observable today [22].

From 1921 to 1945, cane sugar production in DFA increased steadily. The previous social-ecological system was transformed into a monoculture of sugar cane and included in the international trade system [25].

Traditional indigenous common-pool resource institutions that once managed the ecosystem were replaced by the modern state regime.

The transformation of DFA also changed the demographic composition. The labor-intensive nature of cane sugar plantation and industry recruited large numbers of Chinese-descent workers from western Taiwan and China. These Chinese workers, usually called Han people, settled down to form several new settlements around DFA [25]. This explains why today's communities around DFA comprise two main demographics: indigenous people and Han people. These demographics shape the nature of today's collective actions. Some indigenous people were forced to or voluntarily served as sugar industry workers. This contributed to changes in the way indigenous people maintained their livelihoods. Dependence on the sugar industry characterized the economy and social life of communities surrounding DFA until the sugar industry collapsed in 2002.

The sugar plantation also changed the biophysical properties of DFA. Originally a flood plain its high groundwater level hindered maturation of sugarcane. Sponsored by the Japanese government, a drainage system was constructed. Soil improvement measures, such as fertilization and covering new soils from areas outside DFA, were introduced to sustain productivity. These measures were maintained and improved by the next political regime which began in 1945 [25]. The exact ways these measures affect the biophysical properties of DFA today require further study. Continued lowering of groundwater level has raised concerns of local residents and could become a major governance issue in the future.

## 5.4.2 RISE AND FALL OF THE SUGAR PLANTATION: 1945– 2002

After the end of the Second World War in 1945, the central government of the Republic of China (ROC) took over all Japanese assets and state-owned land in Taiwan. The ROC policies further strengthened nationalization [26]. The DFA land and the Salt and Sugar Ltd.'s productive assets were nationalized in 1946 into the Taiwan Sugar Corporation, a public enterprise of the current Ministry of Economic Affairs. After a short period of restructuring due to war damage, the sugar industry had continuously increased production since 1947 [25].

For indigenous people, especially the Karowa community, the change in political regime did not change the nature of oppression exerted by the modern state. Some aspects even deteriorated. During Japanese rule, the Salt and Sugar Ltd. had not fully utilized all DFA land, because of resistance by indigenous people and a subsequent informal agreement with indigenous people in the early stages of Japanese rule. So some indigenous residents still lived and reclaimed approximately 40% of DFA land surrounding the sugar plantation. In 1953, however, the Taiwan Sugar Corporation began to expel residents still living in DFA. Therefore, the indigenous community suffered a second, and more thorough, wave of forced displacement that further deteriorated local livelihoods and community structure [22].

Throughout the 1950s, sugar became one of the Taiwan's most important export industries. Exports accounted for 41.1%–79.8% of Taiwan's total export value. After 1960, rapid industrial development meant sugar exports gradually played a smaller role in economy, even though they still made huge financial contributions to the government and its employees. Sugar production in DFA increased to a peak in 1980. Since the 1980s, the collapse of sugar prices in the international market and the rise of domestic production costs have seriously affected sugar production. Taiwan's membership into the World Trade Organization, valid since 2001, was the last straw. By 2002, sugar production in DFA completely ceased. The land was abandoned from cultivation [25].

With the rise and fall of the sugar industry, the population of Han people in surrounding communities experienced a similar trend. The Daho community, a typical settlement of sugar industry workers, had 6000 people during the peak, but only a few hundred residents live there today. Migration of young people to urban areas caused a huge shortage of human capital, which severely limits governance capacity. This pattern is true for every community neighboring DFA.

Under ROC's authoritarian political regime, before the mid-1980s, local DFA communities obediently followed the jurisdictional and institutional arrangements. During 1980s, however, the rise of the democratic and social movements motivated some indigenous intellectuals to initiate the

"Return My Land Movement" in which indigenous tribes began to claim traditional territories [27]. Inspired by the movement and understanding to the history of his own community, a young man of the Karowa community, Anaw Lo'oh Pacidal, begun to organize at the end of the 1980s, initiating a series of protests throughout the 1990s [8]. Lack of a friendly institutional framework or support from broad social networks prevented progress in his kinship-based protest initiative.

## 5.4.3 FORESTATION: 2003 TO PRESENT

Evolution of the DFA social-ecological system in the current post-sugarcane phase is deeply affected by changes in: (1) forestation policy; (2) the institutional framework regarding indigenous affairs; (3) community development and community forestry policy; and (4) tourism development policy. In the following paragraphs, I describe the impacts of these trends and the interactions among them.

Changes in forestation policy have fundamentally shaped today's ecological system at DFA. Since 2001, the ROC government has promoted the "Afforestation Policy in the Plain Area". Policy objectives include using agricultural land released by accession to WTO to provide ecosystem services, such as environmental aesthetics and carbon sequestration, and increase domestic timber production [28,29]. In 2001, the Taiwan Sugar Corporation owned most of the idle land. Being a public enterprise, it actively cooperated with national policy. Beginning in 2003, the DFA was quickly transformed into a forestation area. In the rush to achieve forestation goals assigned by the ROC government, afforestation in DFA was implemented improperly. Sapling species were chosen and planted depending on availability in the sapling market. Perspectives of stakeholders, especially surrounding communities, were not considered. These problems raised a serious question: could this afforestation project achieve policy objectives in a meaningful way? It is unlikely that such random planting can fulfill any specific policy objective, such as economically efficient use of land, environmental aesthetics, carbon sequestration, or timber production.

The transformation of DFA in 2003 did not consider the controversial nature of land property rights. This issue has increasingly become focus of concern, thanks to the indigenous "Return My Land Movement" and the subsequent changes in institutional consideration of indigenous affairs. A constitutional amendment in 1997, and passing the Indigenous Peoples Basic Law in 2005, recognizes and protects indigenous rights to land and natural resources in traditional territories [30,31]. These laws are the fundamental institutional framework from which many indigenous initiatives lay claim to traditional territories. The Karowa community, led by Anaw Lo'oh Pacidal, resumed to ongoing protest national policies in the first decade of Twenty-first Century. This time, the community formed extensive alliances with indigenous and environmental movement non-governmental organizations to fight for land tenure. Although their actions have not changed the afforestation policy in DFA, they did made a small step toward achieving land rights because of the support of friendlier institutions and bridging organizations. As a symbol of their inherent rights, the Karowa people successfully rented a small piece of land, establishing a resistance stronghold in the middle of the community's original location [8]. Even now, however, the Taiwan Sugar Corporation insists that DFA is its legitimate property.

During 1990s, Taiwan's national policies began to change toward decentralization due to international paradigm shifts and identified faults of the state-centric regime. Since the millennium, every community surrounding DFA was affected by a new national community development policy by the national government, namely the "Community Empowerment". Originally initiated in 1994, this policy aimed to empower local communities and to promote community-based governance, usually by focusing on local cultural, socio-economic and environmental issues [32]. Under the financial and administrative supports of the national government, every community in DFA established its own community development association, sometimes more than one in a community, which is responsible for self-governance.

Almost parallel to the expansion of the community empowerment policy, another new policy also began to exert its influence on these communities. In 2002, influenced by the international trends towards decentralization and participation in environmental governance, the Forestry Bureau initiated the "Community Forestry" policy to promote community-based environmental governance. This policy focused on, but was not limited to, nature conservation and ecotourism [33]. Through the local community development associations, most local communities in DFA have participated in the Community Forestry projects. This participation is directly related to recent tourism development policy.

The tourism development policy is a part of the 12 large-scale economic development plans developed in 2007 by President Ma Ying-Jeou. These plans include extension of the Afforestation Policy in the Plain Area and construction of three large forest parks for tourism development. One of these three parks is DFA, now renamed the Danungdafu Forest Park, and developed to include many additional recreation facilities. The Foresry Bureau took charge of promoting the policy. At the regional level, the bureau's Hualien Branch is responsible for managing the forest park. The park plan and the way the government carried out the plan, offended the Karowa community [34]. As in the past, there was no real participatory and deliberative processes from planning implementation. Indigenous land rights and cultural identity issues were not on the agenda. This time, another indigenous community, Tafalong, joined the protest, inspired by the prevailing indigenous rights movement and its alliance with many indigenous organizations in eastern Taiwan. Karowa, Tafalong and numerous bridging organizations successfully launched a protest during the opening ceremony of the new DFA on 25 May 2011, attracting nationwide attention via media reports.

After the event, public opinion and pressure from their superiors forced the management agency to offer opportunities for participation and collaboration. A meeting was held and attended by all communities surrounding DFA, but it fell apart due to heated debate among communities and the management agency. From 2011 to the present, however, cooperation between the management agency and some communities has strengthened, especially for specific community forestry projects promoting ecotourism. Some communities have also tried to develop a closer inter-community cooperation in the past two years.

Failure of the 2011 meeting can be explained. The management agency was not authorized to deal with fundamental issues such as indigenous land rights and the governance framework. Some people questioned this, concluding that meaningful changes were not possible. To make matters more complicated, there are seriously divergent views among the communities. The temporary alliance between the Karowa and Tafalong communities broke soon after the protest because both claimed ownership of the land tenure and could not reach an agreement.

For communities mainly composed of Han people who were deeply affected by the collapse of sugar industry, the main concerns were improvement of livelihood, not land tenure. These communities include Mafo, Daho, Galiwan, Dahsing, and Fuhsing. They are basically happy with the government's plan to transform forests into a forest park with tourism potential, although they also complain that the government has never provided adequate opportunities for local communities to participate in decision-making processes. Furthermore, these communities do not want open conflict with the government. Under the Community Empowerment and Community Forestry policy, these communities have obtained a number of government projects.

## 5.4.4 SUMMARY OF CROSS-SCALE AND CROSS-LEVEL DYNAMICS

As many social-ecological systems have experienced around the world, transformation were triggered by external drivers related to changes in political jurisdiction at the state level and by economic drivers of international trade at the global level [13]. This change follows introduction of modern state-centric jurisdiction and institutions to replace traditional, local common-pool resources institutions. The history of changes in political jurisdiction fundamentally changed both the ecological and human communities of DFA, the DFA regional social-ecological system. The same global economic and political drivers dominated the evolution of DFA from 1895 to 2002. Both Japanese and ROC governments used DFA land for global market and national economic development, usually at the expense of local communities. Termeer et al. [17] state this is how a typical monocentric governance regime evolves. This regime is characterized by the sole management authority of a central government having only one jurisdictional level ([17], Table 1). Throughout the Twentieth Century,

there was only top-down influence in Taiwan, i.e., the national government. Under the authoritarian regime, local communities were continuously disintegrated and displaced. Until recently, these neither dared nor
had the consciousness and capacity to take bottom-up collective action.
Only one community was the exception, but their weak, kinship-based
action was eventually futile.

At the beginning of the Twenty-first Century, global economic drivers and the monocentric governance regime led to a second transformation. Once again, it was top-down, failing to consider local concerns or
use a true participatory, deliberative decision-making process. This time,
however, the transformation decision was based on both nation-wide economic and environmental aspects. The initiative for carbon sequestration
even cited global concerns, regardless of the feasibility of this argument
for DFA. Probably more critical difference is that two decades of democratic and social movements have produced much stronger nation-wide
networks and a somewhat friendlier institutional framework. This has fos-
tered the emergence of numerous local institutions and bottom-up initiatives. Therefore, two indigenous communities were able to strongly challenge the existing governance regime. Other communities challenged it in
a very moderate way. These challenges were attempts to give feedback on
management issues by upper jurisdictional levels.

The management agency in charge of the Forest Park, the Hualien
Branch of the Forestry Bureau, supervises and guides the five communities implementing community forestry projects. As a result, two-way
communication and feedback is rare between the two jurisdictional levels.
As for the challenges of the Karowa and Tafalong communities, the park
management agency avoids responding to their demands as much as possible. This attitude also applies to the Taiwan Sugar Corporation and its
Hualien branch. As a result, severe hostility and distrust between community and government agencies prevail.

At the community level, inter-community interactions also rarely happen. Local communities are traditionally accustomed to focus on their own
issues. They are rarely aware that, in a multi-scale and multi-level world,
dealing with their own affairs requires multi-party cooperation. This focusing on their own claims usually hinders communication and slows the
building of cooperative relationships among communities. An obvious

example is land claims of the two indigenous communities, which have caused division between the two indigenous communities, and alienation with five other communities. The other five communities focus on livelihood improvement and specific management issues. They are basically willing to work with the park management agency. This attitude makes the two indigenous communities think that other communities are not sympathetic enough with historical oppression suffered by indigenous communities. As a result, trust relationships among communities are either very fragile or non-existent. Hardly any information and knowledge exchange or collective action is observed.

For the lengthy processes of local institution building, all surrounding communities are severely limited by lack of human capital and support networks. This applies to revitalization of traditional institutions in indigenous communities, or emergence of new institutions in other communities. In turn, it limits the ability of initiating intra-community and inter-community collective action.

Finally, the Community Empowerment and Community Forestry policies do show some results, but have some faults. In general, they have good intentions and have made some contribution to local capacity and institution building. Their community-based nature, however, has limited the scope of community actions and vision. Communities were restricted to independently addressing issues within their spatial or institutional jurisdiction. The Forestry Bureau is aware of this problem and has begun to promote inter-community cooperation in the past two years. Real progress remains very limited.

All previously mentioned cross-scale and cross-level dynamics shaped DFA's current governance structure. Figure 3 summarizes actions and interactions among and within different jurisdictional levels. Actions may include command, supervision, guidance, claims making, protest, deliberation, and negotiation.

## 5.5 GOVERNANCE ATTRIBUTES AND CAPACITIES TO MANAGE RESILIENCE OF DFA

In this section I apply the modified framework in Figure 1 to governance attributes and capacity of the DFA.

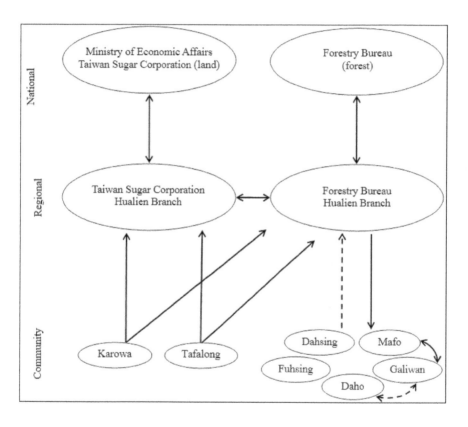

**FIGURE 3:** Current governance structure of DFA at national, regional and community levels. Solid Arrows indicate strong interactions and dashed arrows indicate weak interactions. Arrows indicate direction of interactions. Absence of arrow denotes no interaction.

## 5.5.1 GOVERNANCE ATTRIBUTES

*Participatory and deliberative.* There is a growing consensus that non-state stakeholders need to be involved in the processes of environmental governance. Governance processes include discovering problems, setting agendas, and implementation. Participation helps fulfill the requirement of democratic legitimacy. Participation helps generate critical information for good governance and considers an expanded scope of the knowledge system, diverse interests, and values [11,35]. Deliberation highlights the importance of open communication, discussion, and reflection processes among different stakeholders [11]. The current governance regime of DFA rarely meets the requirements of participation and deliberation. Decision-making for forestation, building the forest park, and formulation of relevant programs, came directly from the policy directives and bureaucratic administrative systems of the national government. It did not include participation and deliberation of non-state actors. Even during implementation, public involvement remains confined within a very limited range. Through community forestry projects, communities have participated in some detailed management issues of the master plan, such as ecological monitoring and ecotourism development. Amendment and discussion of the master plan, however, is still not on the agenda for public participation and deliberation. Communities, non-governmental organizations, and academia were excluded from most decision-making and implementation processes. Currently, no formal or informal mechanisms exist for deliberation that includes non-state participation.

*Polycentric.* The DFA governance structure is characterized by a monocentric authority, not a polycentric institution. The fact that forest and land are managed by separate jurisdictions does not change things, because the Taiwan Sugar Corporation always adheres to national policy. For a small area like DFA, polycentricity may not be an issue of major concern. Polycentric institutions are designed to adapt to spatially heterogenous contexts.

*Multi-layered.* The DFA governance structure currently has three jurisdictional levels: (1) national government; (2) regional management agency of DFA responsible for dutiful implementation of national policies; and (3) community-level institutions still in their infancy. Functionally, it is a two-tier structure with a dominant state and weak community. This is a moderate version of the stereotypical dominance pattern classified by Young [14]. Can this structure deal with scale-dependent challenges and cross-scale interactions [11]? Basically, the answer is no. As shown in Section 4, the national government, its policies, and its agents consider land use options mainly from national or global perspectives. The forestation policy is a typical example. The policy itself, and the way the policy was carried out, overlook the social-ecological settings of DFA. Local communities and academia have asked some fundamental questions: such as whether DFA biophysically suitable for timber production and carbon sequestration and how a democratic government can repeat the actions of the past authoritarian regimes pursuing national goals at the expense of local needs and inherent rights. When the current state government does discover and tries to fix this imbalance, its community-based policies overemphasize on community-level management issues, as do the communities themselves. In some cases, communities have made proposals without considering possible negative externalities. As Berkes [13] has highlighted, states and communities usually define issues from the perspective of their own level, without considering perspectives of other levels. What makes matters worse is the lack of effective cross-level linkages. The state dominates decision-making and implementation. It is reluctant to devolve authority to or receive feedbacks from other levels. This fundamentally limits possible emergence of multi-layered governance. Issues at DFA can rarely be addressed at suitable spatial and institutional levels, nor can they be adequately addressed by appropriate cross-scale and cross-level linkages. Therefore DFA represents a typical example of scale mismatch in which institutional arrangements do not fit the social-ecological system boundary [13,36].

*Accountable.* Accountability, by definition, means the processes and mechanisms through which actors are responsible for providing information and indicating the basis of decision-making and actions to the point of being punished when there is misconduct [37]. This concept can be divided into upward and downward accountability [11]. In the case of DFA, upward accountability exists, as usually happened in other cases. At DFA, local communities are obliged to be accountable to authorities at upper levels, via a wide range of governmental regulations and projects. Other than a few superficial public hearings triggered by indigenous movements and the subsequent public pressure, downward accountability is virtually non-existent.

*Just.* Given the governance attributes of DFA, it is reasonable to expect unjust distribution of benefits and involuntary risks to last for a very long time. The forests of DFA do generate some public goods, common-pool resources, and economic benefits, but local communities, especially the historically oppressed and currently vulnerable indigenous groups, bear most of the structure- and policy-induced costs. These people have lost their land tenure, their use rights, and their minimal rights of being formally included in the governance regime.

*Networked.* The attribute networked refers to whether and how different stakeholders come together to deal with governance issues. Recent research [19,38] has shown that social network patterns exert important influences on governance processes and outcomes. Bodin and Crona [19] list five main network characteristics affecting governance: number of ties, degree of network cohesion, subgroup inter-linkages, network centralization, and actor centrality. According to these characteristics, DFA represents a highly centralized network pattern, in which government agencies are the central actors loosely connected to the relatively isolated local actors (Figure 3). Although high network centrality may help solve simple problems, it may not be appropriate for dealing with complexities common in social-ecological systems, such as DFA. On the other hand, local actors are isolated and weak. There are few social ties among the isolated communities. Between the two clearly distinguishable subgroups (indigenous and Han) there is poor network cohesion. There are hardly any bridging ties for subgroup inter-linkage. This lack of ties, cohesion and inter-linkage, is generally considered to have negative effects on the formation of trust relationships, informational and intellectual exchange, collective action, and good governance. These negative effects are clearly shown in Subsection 4.4.

### 5.5.2 CAPACITY TO MANAGE RESILIENCE

To manage social-ecological systems for resilience, a number of capacity dimensions are needed as Lebel et al. [11] and Figure 1 suggest. In the following analysis of the capacities of the governance regime at DFA, I follow the interpretations and terminology of Lebel et al. [11], except for the addition of social capital and leadership.

*Knowledge.* Knowledge is the capacity "to combine and integrate different forms of knowledge" [11], which represent a precondition for the conservation of biodiversity and cultural diversity [39]. At DFA, indigenous traditional knowledge has never been acknowledged by or included in the governance regime. Indigenous communities have used their food forest tradition and memories from the pre-state era to propose improvement ideas for combined forest and land use options that recover both biological and cultural diversity. Other local communities have also provided proposals based on their local knowledge of ecosystems and water scarcity. They, too, have not received attention from government authorities. The modern academic knowledge system usually enjoys an authoritative status in Taiwan, but only a small part of it (i.e., forestry) has played a role in decision-making and implementation processes. In recent years, some academic research projects have included multi-disciplinary perspectives. Right now, it is hard to expect today's governance regime, especially the central actors, to not only recognize the importance of integrating different knowledge systems, but also to take action to include them into governance processes. Only when fundamental governance attributes, such as participation and deliberation, improve can the flaws in knowledge capacity be corrected.

*Diversity.* The ignorance and unwillingness of knowledge seriously affects "diversity", the capacity "to maintain ecological and social diversity" [11]. Indigenous knowledge system, institutions, and cultural identity have suffered serious loss. If this trend remains unchanged, eventually indigenous communities will be lost as a vital source of renewal and reorganization in the DFA system. Despite hasty decision-making and implementation processes, forestation policies have contributed to the recovery of species diversity. The monoculture nature of forestation and the governance processes that ignore different knowledge systems, will eventually limit the extent to which biological diversity can recover.

*Uncertainties.* Uncertainties represents the capacity "to anticipate and cope with uncertainties and surprises" [11]. This needs an attitude open to accept change and the ability to learn by doing. Government agencies tend to maintain the status quo of national policies. They do not tend to face the possibility of change or accept views from different knowledge systems. They especially do not develop the double- or triple-loop learning central to adaptive governance [40]. A few experiments at the level of single-loop learning have been carried out via community-based projects. The premise of these experiments is that the existing forest park and governance structure will remain unchallenged.

*Thresholds.* Thresholds means the capacity "to detect and navigate hard-to-reverse thresholds" [11]. The existing governance regime, through cross-level interaction, discovered a few possible thresholds, such as over-exploitation of some wildlife species as a result of poaching. The regime, however, remains insensitive and unresponsive to upcoming thresholds, such as collapse of the indigenous knowledge system, alteration of land tenure by legislation of the Indigenous Land and Ocean Act, and planning for a new freeway that will run through the DFA.

*Scale and fit.* The closely related scale and fit represent the capacity "to engage effectively with and handle multiple- and cross-scale dynamics" and the capacity "to design institutions which fit diverse social and eco-logical contexts", respectively [11]. Using the scales and levels shown in Table 1, the current governance regime in DFA has fundamental flaws, as indicated in Subsection 5.1. Government authorities are mainly concerned about national and global spatial scales, national jurisdictional scales, laws and regulations (institutional scale), and general knowledge. Local com-munities, whether because of their inherent nature or because of commu-nity-based governance projects, focus on patch spatial scales, community jurisdictional scales, local institutional scales, kin or community network scale, and specific knowledge. Between national and local communities, there are no intermediate or bridging institutions and few effective cross-scale and -level linkages to fit the DFA social-ecological system.

*Social capital.* Social capital is an important capacity contributing to sound environmental governance [41,42]. Specifically, it includes bonding and bridging networks, and trust among individuals and groups of people [19]. Currently, each local community in DFA has its own bonding net-works, mainly kinship-based. Among communities, bonding relationships are very weak (see Subsection 4.4). Two indigenous communities have developed their own bridging networks with non-governmental organizations, but the other communities are very weakly connected. As mentioned in Subsection 4.4, hostility and distrust prevail among the two indigenous communities and government agencies. Alienation is obvious between indigenous and Han communities. The fragile state of social capital has historical and governance roots.

*Leadership.* Good leadership with human skills and agency are needed for adaptation and transformation towards sustainability [43,44,45]. Westley et al. [43] list the fundamental skills contributing to successful ecosystem stewardship: "facilitating knowledge building and utilization; vision building; developing social networks; building trust, legitimacy, and social capital; facilitating/developing (social) innovations; preparation and mobilization for change; recognize or create and seize windows of opportunity; identifying and communicating opportunities for 'small wins'; and facilitate conflict resolution and negotiations". In DFA, a few community leaders have some of these skills, but they have almost exclusively been applied to community-level issues. So far, no leaders have emerged with the vision and skills for transformation at the DFA or higher system perspective. This includes governmental officials. Some bridging organizations and individuals are involved in governance of DFA, but mainly in collaborative roles. But still, none has lead or shown a willingness to lead in transformation.

## 5.6 CONCLUSIONS AND POLICY IMPLICATIONS

Resilience thinking has strongly influenced how people understand and pursue sustainably linked social-ecological systems. Resilience thinking highlights the need to build capacity and manage general system properties in a complex, constantly changing world. I addressed social-ecological system governance, and explored associations between cross-scale and cross-level dynamics, governance attributes, and capacities to manage resilience in the social-ecological system of DFA. Based on previous studies [11,15], I propose a modified analytical framework that permits analysis of associations among essential governance attributes and the capacities to manage resilience. There are two main conclusions from this Taiwanese example of a social-ecological system: the modified framework was useful and it helped identify flaws in current governance.

The modified framework, when supported by an in-depth investigation into cross-scale and cross-level dynamics, can be useful for understanding how current capacities for managing resilience are related to fundamental governance attributes. Generally, capacities to manage resilience are greatly influenced by governance attributes. Governance attributes are the outcomes of the long-term dynamics of history, politics, ecology, society and economy. Thus, at DFA, there is an obvious path-dependence phenomenon.

The modified framework helped identify fundamental flaws of the current governance regime, including key issues needing to be addressed. Overall, the governance regime of DFA was characterized as (1) non-participatory and non-deliberative; (2) monocentric with a two-tier structure of dominant state and weak communities; (3) upward accountability and unjust; and (4) very loose networks. These attributes help explain the serious defects in DFA governance. The current governance regime has extremely limited capacities in knowledge, diversity, uncertainties, thresholds, social capital, and leadership. The current governance regime shows a mismatch in scale and fit dimensions. According to the classification of Termeer et al. [17], this governance regime is a traditional monocentric regime. Although it has recently begun to attempt to include the limited contents of multilevel governance, it is still far away from genuine multilevel governance. In particular, hardly any of the actors, whether government authorities or local communities, have begun to learn and recognize the importance of adaptive governance. As Termeer et al. [17] emphasize, scale-sensitive governance regimes are needed for adaptation to complex social-ecological conditions.

What can be done to improve current governance attributes and capacities of the studied governance regime? Analysis in Subsection 5.2 indicates a critical starting point: government and local communities should recognize that the multi-scale and multi-level interactive world means their individual and separate capacities are far from sufficient to deal with complex challenges. This system needs to navigate toward a governance regime that is more participatory, deliberative, multi-layered, accountable, just, and networked. This can be done by developing an intermediate institution to coordinate cross-scale and cross-level interactions in ways that may more adequately fit the DFA social-ecological system. Numerous empirical examples around the world have shown the value of these intermediate institutions [13,46,47,48]. In the DFA, the government should upgrade its community-based policy to a regional social-ecological system governance policy that encourages bottom-up participation of local communities in governance. As part of its top down action, the government should admit to the historically unjust nature of current land tenure and dominant governmental power. By creating a reconciliation atmosphere, the government can facilitate subsequent negotiation, collaboration, and transformation. Equally important, the government should eventually devolve part of its authority to other governance levels, depending on the nature of the governance issue. This awareness, consensus, and external institutions should be cultivated to create a friendly environment from which innovative and adaptive governance initiatives can emerge to function successfully.

## REFERENCES

1. Brundtland, G.H. Report of the World Commission on Environment and Development—Our Common Future; United Nations General Assembly: New York, NY, USA, 1987.
2. Anderies, J.M.; Folke, C.; Walker, B.; Ostrom, E. Aligning key concepts for global change policy: Robustness, resilience, and sustainability. Ecol. Soc. 2013, 18, Article 8.
3. Holling, C.S. Understanding the complexity of economic, ecological, and social systems. Ecosystems 2001, 4, 390–405.
4. Folke, C.; Carpenter, S.R.; Walker, B.; Scheffer, M.; Chapin, T.; Rockström, J. Resilience thinking: Integrating resilience, adaptability and transformability. Ecol. Soc. 2010, 15, Article 20.
5. Chapin, F.S., III; Folke, C.; Kofinas, G.P. A framework for understanding change. In Principles of Ecosystem Stewardship: Resilience-Based Natural Resource Management in a Changing World, 1st ed.; Chapin, F.S., III, Kofinas, G.P., Folke, C., Eds.; Springer: New York, NY, USA, 2009; pp. 3–28.
6. Walker, B.; Holling, C.S.; Carpenter, S.R.; Kinzig, A. Resilience, adaptability and transformability in social-ecological systems. Ecol. Soc. 2004, 9, Article 5.
7. Chapin, F.S., III; Carpenter, S.R.; Kofinas, G.P.; Folke, C.; Abel, N.; Clark, W.C.; Olsson, P.; Smith, D.M.S.; Walker, B.; Young, O.R.; et al. Ecosystem stewardship: Sustainability strategies for a rapidly changing planet. Trends Ecol. Evol. 2009, 25, 241–249.
8. Hwaung, Y.-H. Multiple Boundaries as Palimpsests: The Landscapes of Native Title in Taiwan. Ph.D. Thesis, National Dong Hwa University, Hualien, Taiwan, 2014. (In Chinese).

9.  Lemos, M.C.; Agrawal, A. Environmental governance. Annu. Rev. Environ. Resour. 2006, 31, 297–325.
10. Young, O.R. The effectiveness of international institutions: Hard cases and critical variables. In Governance without Government: Order and Change in World Politics; Rosenau, J.N., Czempiel, E.-O., Eds.; Cambridge University Press: Cambridge, UK, 1992; pp. 160–196.
11. Lebel, L.; Anderies, J.M.; Campbell, B.; Folke, C.; Hatfield-Dodds, S.; Hughes, T.P.; Wilson, J. Governance and the capacity to manage resilience in regional social-ecological systems. Ecol. Soc. 2006, 11, Article 19.
12. Cash, D.W.; Adger, W.N.; Berkes, F.; Garden, P.; Lebel, L.; Olsson, P.; Pritchard, L.; Young, O. Scale and cross-scale dynamics: Governance and information in a multi-level world. Ecol. Soc. 2006, 11, Article 8.
13. Berkes, F. From community-based resource management to complex systems: The scale issue and marine commons. Ecol. Soc. 2006, 11, Article 45.
14. Young, O. Vertical interplay among scale-dependent environmental and resource regimes. Ecol. Soc. 2006, 11, Article 27.
15. Armitage, D. Governance and the commons in a multi-level world. Int. J. Commons 2008, 2, 7–32.
16. Gibson, C.C.; Ostrom, E.; Ahn, T.K. The concept of scale and the human dimensions of global change: A survey. Ecol. Econ. 2000, 32, 217–239.
17. Termeer, C.J.A.M.; Dewulf, A.; van Lieshout, M. Disentangling scale approaches in governance research: Comparing monocentric, multilevel, and adaptive governance. Ecol. Soc. 2010, 15, Article 29.
18. Wilson, J.A. Matching social and ecological systems in complex ocean fisheries. Ecol. Soc. 2006, 11, Article 9.
19. Bodin, Ö.; Crona, B.I. The role of social networks in natural resource governance: What relational patterns make a difference? Glob. Environ. Chang. 2009, 19, 366–374.
20. Resilience Alliance. Assessing Resilience in Social-ecological Systems: Workbook for Practitioners, Version 2.0. 2010. Available online: http://www.resalliance.org/3871.php (accessed on 13 August 2014).
21. Chang, T.-Y.; Tsai, B.-W. Indigenous Traditional Territory: Research Report, 1st ed.; Council of Indigenous People, Executive Yuan: Taipei, Taiwan, 2003. (In Chinese).
22. Hwaung, Y.-H. Hometown of Others: On Displacement and Autonomy Movement of Karowa Indigenous People from the Perspective of Space Hegemony. Master's Thesis, National Dong Hwa University, Hualien, Taiwan, 2003. (In Chinese).
23. Shizue, F. Governing Indigenous People: The Plan of Governing Taiwan by Japan, 1st ed.; WenYingTang: Taipei, Taiwan, 1997. (In Chinese).
24. Lin, S.-C. Land and Settlement Development in Central East Rift Valley: 1800–1945. Master's Thesis, National Taiwan Normal University, Taipei, Taiwan, 1995. (In Chinese).
25. Chung, S.-H. A History of the Development of Cane Sugar Industry in Hualien: 1899–2002, 1st ed.; East Taiwan Research Association: Taitung, Taiwan, 2009. (In Chinese).
26. Liu, N.-Y. Notes on Taiwan's Land Reform; Literature Institute of Taiwan Province: Nantou, Taiwan, 1989. (In Chinese).

27. Parod, I., Ed.; Archives of the Taiwan Indigenous Movement; Council of Indigenous People and Institute of National History: Taipei, Taiwan, 2008. (In Chinese).

28. Lin, K.-C. The analysis of the afforestation policy in the plain area. Taiwan. Agr. Econ. Rev. 2003, 8, 111–140, (In Chinese).

29. Chen, A.-H.; Chen, L.-M. Promoting the afforestation policy in the plain area. Taiwan For. J. 2002, 28, 18–21, (In Chinese).

30. Yapasuyongu Poiconu. United Nations declaration on the rights of indigenous peoples and protection of Taiwan indigenous rights. J. Taiwan. Ind. Stud. 2007, 2, 141–168, (In Chinese).

31. Huang, S.-M. Globalization and the change and status quo of fundamental indigenous policy in Taiwan. In Government Policy and Social Development among Taiwanese Indigenous Peoples, 2nd ed.; Huang, S.-M., Chang, Y.-H., Eds.; Institute of Ethnology, Academia Sinica: Taipei, Taiwan, 2010; pp. 15–50, (In Chinese).

32. Chen, L.-C. Recent development of community empowering in Taiwan. J. Hous. Stud. 2000, 9, 61–77, (In Chinese).

33. Lu, D.-J. Indigenous people and community forestry. Taiwan J. For. Sci. 2009, 16, 28–30, (In Chinese).

34. Hwaung, Y.-H. Indigenous traditional territory? National territory tradition? Enlightenment of a dialogue. Cul. Stud. Bimonth. 2012, 132, 69–87, (In Chinese).

35. Bulkeley, H.; Mol, A.P.J. Participation and environmental governance: Consensus, ambivalence, and debate. Environ. Values 2003, 12, 143–154.

36. Cumming, C.S.; Cumming, D.H.M.; Redman, C.L. Scale mismatches in social-ecological systems: Causes, consequences, and solutions. Ecol. Soc. 2006, 11, Article 14.

37. Schedler, A. Conceptualizing accountability. In The Self-Restraining State: Power and Accountability in New Democracies; Schedler, A., Diamond, L., Plattner, M.F., Eds.; Lynne Rienner Publishers: London, UK, 1999; pp. 13–28.

38. Bodin, Ö.; Crona, B.I.; Ernston, H. Social networks in natural resource management: What is there to learn from a structural perspective. Ecol. Soc. 2006, 11, Article 2.

39. Rozzi, R. Biocultural ethics: From biocultural homogenization toward biocultural conservation. In Linking Ecology and Ethics for a Changing World: Values, Philosophy, and Action; Rozzi, R., Pickett, S.T.A., Palmer, C., Armesto, J.J., Callicott, J.B., Eds.; Springer: New York, NY, USA, 2013; pp. 9–32.

40. Kofinas, G.P. Adaptive co-management in social-ecological governance. In Principles of Ecosystem Stewardship: Resilience-Based Natural Resource Management in a Changing World, 1st ed.; Chapin, F.S., III, Kofinas, G.P., Folke, C., Eds.; Springer: New York, NY, USA, 2009; pp. 77–101.

41. Pretty, J. Social capital and the collective management of resources. Science 2003, 302, 1912–1914.

42. Berkes, F. Evolution of co-management: Role of knowledge generation, bridging organizations and social learning. J. Environ. Manag. 2009, 90, 1692–1702.

43. Westley, F.R.; Tjornbo, O.; Schultz, L.; Olsson, P.; Folke, K.; Crona, B.; Bodin, Ö. A theory of transformative agency in linked social-ecological systems. Ecol. Soc. 2013, 18, Article 27.

44. Moore, M.-L.; Westley, F.R. Surmountable chasms: Networks and social innovation for resilient systems. Ecol. Soc. 2011, 16, Article 5.

45. Olsson, P.; Folke, K.; Galaz, V.; Hahn, T.; Schultz, L. Enhancing the fit through adaptive co-management: Creating and maintaining bridging functions for matching scales in the Kristianstads Vattenrike Biosphere Reserve, Sweden. Ecol. Soc. 2007, 12, Article 28.
46. Olsson, P.; Folke, C.; Hahn, T. Social-ecological transformation for ecosystem management: The development of adaptive co-management of a wetland landscape in southern Sweden. Ecol. Soc. 2004, 9, Article 2.
47. Borgström, P.; Elmqvist, T.; Angelstam, P.; Alfsen-Norodom, C. Scale mismatches in management of urban landscapes. Ecol. Soc. 2006, 11, Article 16.
48. Olsson, P.; Folke, C.; Hughes, T.P. Navigating the transition to ecosystem-based management of the Great Barrier Reef, Australia. Proc. Natl. Acad. Sci. USA 2008, 105, 9489–9494.

# Towards Enhanced Resilience in City Design: A Proposition

ROB ROGGEMA

## 6.1 INTRODUCTION

Resilience is defined as "the capacity of a system to absorb disturbance and reorganize while undergoing change so as to still retain essentially the same function, structure, identity and feedbacks" [1,2]. It includes the ability to learn from disturbance [3]). A core question of resilience theory is how resilience in a system could be increased. In the context of this paper, the systems in question are urban-ecological and socio-ecological systems; psychological resilience (e.g., [4,5]); and the physical properties of material [6] or resilience engineering [7], are beyond the scope of this article. The theoretical framework relating to resilience (e.g., [2]) and resilient cities (e.g., [8]) focuses strongly on the mechanics of the system or the city, but to date, this has not resulted in more resilient cities. Concrete, practical directions on how to build a resilient city are often lacking, even when the metabolism of the city is known in detail (e.g., [9]).

*Towards Enhanced Resilience in City Design: A Proposition.* © *Roggema R.* Land *3,2 (2014), doi:10.3390/land3020460. Licensed under a Creative Commons Attribution 3.0 Unported License, http://creativecommons.org/licenses/by/3.0/.*

One key element of resilience theory is the capability of a system to bounce back, or rearrange the functions or elements of the ecosystem according to the requirements imposed by the environment, or external shocks. The city, by its built nature, has a limited capacity to rearrange its built objects, such as buildings and infrastructure. In this article, this is seen as the key obstacle to increasing resilience. A further key aspect of resilience is that the system should anticipate future impacts and potential change. This implies a design requirement to create a plan for building cities, which are flexible whenever required in the future.

The role herein for design, urban design, landscape architecture and urban planning, is up till the present day limited, a reason to propose a "Dismantable City", which, in this article is defined as: "A city with urban elements or objects, such as buildings, and (parts of) streets, sewage or other infrastructure, which can be freely (re)arranged. In this city, both functional and physical parts are seen as temporary, allowed to change their location and/or function". The word dismantable does not exist in the English language. Dismantle does exist, defined as: take (a machine or structure) to pieces. We use dismantable to express the potential to take (the city) apart. Hypothetically, the mobility of elements enhances the flexibility of spatial configurations, hence, increases resilience. Especially when huge (climate) impacts are expected, the capability to change is paramount. Within the context of this article, social aspects are not considered.

The argument will be made that principles drawn from the fields of self-organization and complexity need to be used in urban design in order for urban systems to be genuinely more resilient. Furthermore, it will be argued that the concept of the Dismantable City offers a concrete method for this application of self-organization and complexity in urban design, and, hence, that the Dismantable City offers a novel way of designing urban and social systems, which will increase those systems resilience.

Section 2 will define resilience and adaptive capacity and will explain their foundations in existing research. It will also make clear how aspects of self-organization increase the resilience of a complex system. Section 3 presents a city as a complex (eco)system, which, as such, would benefit from self-organization when a higher degree of resilience is desirable. In Section 4 the metabolism that results from current planning practice is discussed, with its effects on (the lack of) resilience. In Section 5 the Dismantable City is offered as a self-organizing alternative to current urban design, applying the best principles of self-organization to urban elements. Three examples of the Dismantable City from existing planning practice are provided in Section 6. Finally, constraints to implementation of the Dismantable City from current planning practice are discussed.

## 6.2 ADAPTIVE CAPACITY

Resilience, or adaptive capacity, is one of three properties of the adaptive cycle and is defined as "a measure of its vulnerability to unexpected or unpredictable shocks... (that) can be thought of as the opposite of the vulnerability of the system" [10]. An adaptive cycle alternates between long periods of aggregation and transformation of resources and shorter periods that create opportunities for innovation. This cycle is seen as fundamental to understanding complex systems, from cells to ecosystems to societies. For ecosystem and socio-ecological system dynamics that can be represented by an adaptive cycle, four distinct phases have been identified:

1. Growth or exploitation (r)
2. Conservation (K)
3. Collapse or release ($\Omega$)
4. Reorganization ($\alpha$)

The adaptive cycle exhibits two major phases (or transitions). The first, often referred to as the fore-loop, from r to K, is the slow, incremental phase of growth and accumulation. The second, referred to as the back-loop, from $\Omega$ to $\alpha$, is the rapid phase of reorganization leading to renewal [11]. The $\alpha$-phase is especially important for the resilience of the system [12]:

*"As the phases of the adaptive cycle proceed, a system's ecological resilience expands and contracts. The conditions that occasionally foster novelty and experiment occur during periods in the back loop of the cycle, when connectedness, or controllability, is low and resilience is high (that is, during the $\alpha$-phase).*

*The low connectedness, or weak control, permits novel re-assortments of elements that were previously tightly connected to others in isolated sets of interactions. The high resilience allows tests of those novel combinations because the system-wide costs of failure are low. The result is the condition needed for creative experimentation."*

Generally, resilience is seen as a property of socio-ecological systems, allowing the system to bounce back and recover from an external shock. Adaptability is the capacity of actors to influence resilience [1]. The collective capacity to manage resilience determines whether actors can successfully avoid crossing into an undesirable system regime or return to a desirable one. Actors may use four ways to influence resilience:

- Move thresholds;
- Make the threshold more difficult to reach;
- Move the system away from the threshold;
- Avoid loss of resilience by managing cross-scale interactions.

The question is, however, whether these principles are applicable to city design. To a certain extent they are, for instance in the way water, nature, and social structures are designed. It is beneficial for the sustainability of the city to include nature, cleaner water and inclusive social environments in designs. However, do these features still function in disaster-like circumstances, or when the risk landscape changes from predictable changes towards uncertain and unprecedented climate events [12,13]?

The city needs to be able to constantly change its shape, its functionality and its urban fabric under the influence of external shocks and to relocate urban objects when necessary [14]. Four key properties of resilience, enhancing flexibility and adaptability of built environments and communities, can be defined [15]. These are the need to: create redundancy, which provides free space where new functions may fit, increase diversity multiplying the number of different objects, transform towards a larger modularity where modules can be easily disconnected from others, and allow for feedback through which objects can mutually interact, develop self-organization, and find optimal spots. These properties form a preliminary set of guiding principles for building resilience into city design.

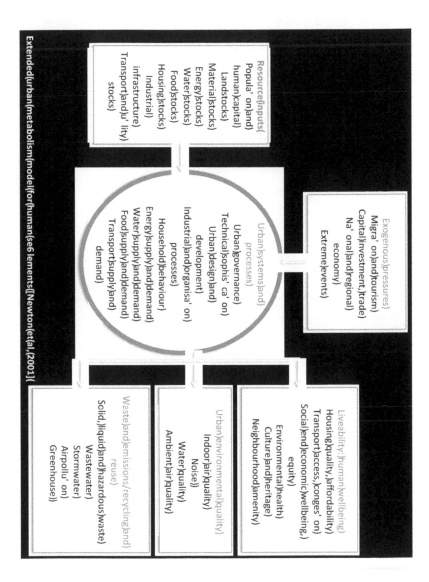

**FIGURE 1:** Extended urban metabolism model (adapted from [51]).

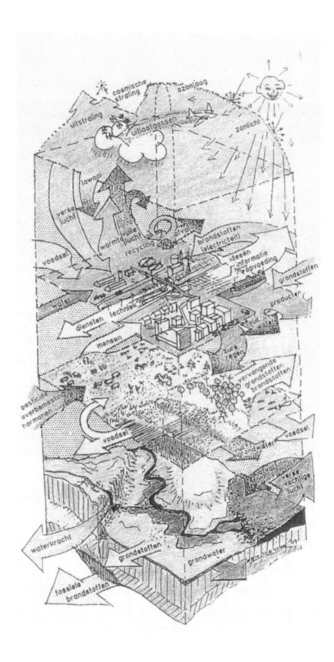

**FIGURE 2:** Combining the layer components of the urban system with in- and out-going flows: Kristinsson's environmental system [54].

According to Walker et al. [1], the core element in increasing adaptability is the opportunity of agents to manage resilience. These agents, which we call urban objects when talking about cities, need space to move around and be flexible as to where and how they habituate. This implies a "comfortable chaos" [16] in which too much organization can be a bad thing. Comfortable chaos emerges when the city regenerates (α-phase) and creates the space for adjustments, repair, change, and innovation. This happens in times of crisis, when there is a certain mess and room for mistakes.

## 6.3 THE CITY AS AN (ECO)SYSTEM

In order to dig deeper into the resilience of the city, the city needs to be viewed as a system. Complexity theory is very helpful here. Prigogine and Stengers, who studied non-linear dynamic systems [17,18] and Kauffman, who studied self-organization [19], are considered the founding fathers of Complexity Theory. The first researchers to adopt their ideas were natural scientists and physicists. As computers became more powerful, however, their ideas also found their way to researchers working on computer simulations of social systems and artificial life [20]. Nowadays, many scholars use features of Complexity Theory in studies of a broad range of natural and human systems. Complexity Theory is seen as a science enabling the gap between social and natural sciences to be bridged [17,21,22,23,24]. Complexity Science has developed from studying closed systems to the study of open systems, including real-life situations, in a wide range of social sciences [25,26,27,28,29,30,31,32,33,34,35,36]. The sudden unexpected change from one attractor into another, as well as aspects of dealing with the uncertainty of possible unexpected change, have been thoroughly examined. Although spatial planning is deeply rooted in a control paradigm [37], the outcome of a planning process can differ greatly from the intended outcome. Hence, the results can be highly surprising and spatial planning intrinsically has to deal with uncertainty and fuzziness [38,39,40,41].

Odum used an analogy of electrical energy networks to model the energy flow pathways of ecosystems [42]. This played a significant role in developing his approach to systems and is recognised as one of the earliest instances of systems ecology. Thinking in sources, production, consumption and losses inspired many to use this language in thinking about the city as an ecosystem. It was especially useful when the cities' environmental problems needed to be solved.

The city is conceived as a dynamic and "double" complex ecosystem [43]. The social, economic and cultural systems cannot escape the rules of abiotic and biotic nature. Guidelines for action will have to be geared to these rules [44].

Combining this idea with the concept of urban metabolism [45], a method to analyse cities and communities through the quantification of inputs (water, food, and fuel), outputs (sewage, solid refuse, and air pollutants), and tracking their respective transformations and flows, led to the extended urban metabolism model [46], which has been further elaborated by many [8,47,48,49,50,51,52]. In this model (see Figure 1), the inputs (flows, sources, stock) are defined that enter the urban system. Under the influence of many exogenous factors, these are transformed, producing sources for use in the city and leading to certain environmental qualities, waste and a certain liveability. To improve sustainability the ingoing flows need to be reduced, the internal processes improved, waste reduced and both environmental quality and liveability improved.

This model of a system with input, internal processing and output is then applied to a city. Odum's scheme is adapted to a 3D image with in- and outgoing flows and detailed with environmental parameters, divided over several layers (derived from [53]) (Figure 2, [54]).

This gives us the parameters to measure or assess the sustainability of a city, but can it also enhance the resilience? In order to achieve this, the city must be capable of regenerating constantly, of reconfiguring its layout and of relocating urban objects. This must be done in a way that reduces the need for external sources, can be adjusted to changing exogenous pressures, can produce efficiently, can reuse and harvest what is available in the city, and can provide a liveable, clean environment that produces no or minimal waste. The city can only perform in this way when urban objects have the opportunity to self-organize, defined as "the potential to spontaneously and unpredictably develop new forms and structures by itself out of chaos" [55], i.e., to be able to "freely" move around and find new locations in interaction with others and the exogenous environment. This can only happen when these objects are dismantable and reorganizable in a permanent search for the optimal spots in each circumstance. In many cases the reconfiguration of the urban object itself may be sufficient, for instance when heat waves strike, or a blizzard hurts the city, but not in all circumstances this is sufficient. In case of a flood or bushfire, events with a clear geographical impact, adjusting the urban objects only could not prevent them from vanishing. Here, mobility is key.

This is the Dismantable City: where urban objects can be reconfigured and, when necessary, be taken apart, reused, and relocated.

## 6.4 SUPPORT OF CURRENT PLANNING METHODS FOR ENHANCING RESILIENCE

The metabolism that results from current spatial planning practice visualized as an (eco)system with flows that enter and leave the system, is not very sustainable. These types of systems generally consume landscape, produce money, and leave an inflexible mono-functional urban environment. The result is a static urban environment far from resources or the central city, consisting of unattached half-acre blocks, dependent for jobs, food, water, and energy on incoming sources and leaving the environment with waste and pollution. In other words, the urban metabolism of current city planning is not sustainable, nor does it enhance resilience. Land-use in cities uses resources, in the form of food, fossil fuels, water, and energy. The majority of resources used are finite and only marginally renewable. Further metabolism thinking has been focused on a linear throughput of resources through the city, aiming to use finite resources and flows more efficiently, i.e., with fewer leftover materials. Recently, circular metabolism is promoted as a form of city development in which these leftover materials, preferable to renewable resources, are recycled and reused back into the urban system. Given the idea that a sustainable (eco)system is resilient by definition, the problem is not the resilience of the system itself, but the translation of resilience concepts in urban planning and design. Therefore, the following discussion focuses on the way planning approaches support a more resilient development (or not). It implies that a linear planning methodology, if it would embrace a circular metabolism, at least should become circular itself, set aside the option of being non-linear. In the recent planning theory discourse, there are clear signals to replace mono-rationality, for the ease of the argument equated to linearity, with more complex and self-organizing methods of planning, which take into account uncertainty, flexibility and emergence.

Planning theory supports, though presently in a very limited way, a move away from a non-innovative state of mono-rationality [56], which generally conceives of only one way of "good" planning [57] that is dominated by the government, in close association with developers. In order to establish planning without tightening and dictating regulations, current planning practice must be replaced by an alternative, that has been variously described as: post-anarchistic, or autonomous planning creating a disordered order of "becoming" spaces [58]; planning from "outside inward" [59]; informal, insurgent, planning [60]; planning by surprise, making use of coincidental opportunities [61]; or poly-rational unsafe planning [56]. This alternative looms when the fundamental properties of Western planning mono-rationality, namely "playing by the rules", "repeat habitual prior experiences", and "creating a non-innovative status quo" [56], all withstanding self-organization and indirectly resilience, are left behind. Then, "liquid, turbulent or even wild boundaries of both planning thought and spatial territory can occur—literally, to do 'it' without the safety of a condom!" [56]. This is a planning practice that takes risks, accommodates difference and encourages the new and creative. In this Dismantable City, physical and functional mobility will be possible and a city is built that is constantly adaptable and resilient. However, before replacing the existing planning system, no matter how inaccurate, an alternative must be studied further or, for the time being, both planning approaches should co-exist.

## 6.5 A DISMANTABLE CITY?

The proposition of a Dismantable City is new. The following principles have been identified:

1.  Dismantle building components in order to reuse them in other configurations again.

2.  Create urban objects that are detached from their environment.
3.  The identification of specific zones where mobile elements can relocate in periods of change/turbulence.
4.  Developing plug-in infrastructure where mobile urban objects can attach.

There have been only a few movements that show similarities with the Dismantable City concept, such as the Plug-In-City conceived by the Archigram group, consisting of a framework only to be occupied by plug-in objects [62,63], or Superuse [64] concepts, which "superuse" available flows and resources and connect them into urban ecosystems [65]. The idea is derived from building practice in which the option to disassemble building elements, deconstruct the building and reuse the parts in another building has been long studied. Deconstruction focuses on giving the materials within a building a new life once the building, as a whole, can no longer continue, by "carefully dismantling a building in order to salvage components for reuse and recycling" [66]. Taking it one step further, the adjustability of building components can be increased by programmatic labelling and tagging of building elements, enabling buildings to customize temporary desires or changing demands, so that the building elements perform swarming behaviour [67,68,69]. At a larger scale, modular buildings have similar ambitions. For instance, the Habitat 67 project in Montreal (Figure 3) is a residential structure consisting of separate, functional apartments that can be put together in a variety of ways. As people move in or out, units can be reconfigured as desired [70]; although it must be mentioned that, in practice, this has not happened very often.

In areas under threat of flooding propositions of floating cities are well known, for instance for the London Theems Gateway (e.g., [71]) and in cities, such as Bangkok, Hanoi, or Djakarta. More common examples can be found in the USA, where living on House Boats is not very unusual, or the recent developments in the new urban development of IJburg, Amsterdam, where a substantive section of the new housing is floating on the water. Experimental landscape design for a region under threat of flooding, where housing follows and moves towards places that are out of the danger zone is the design for the Floodable Landscape in the Eemsdelta region in the Netherlands [72].

**FIGURE 3:** The Habitat 67 project in Montreal (Picture: Ifte Ahmed).

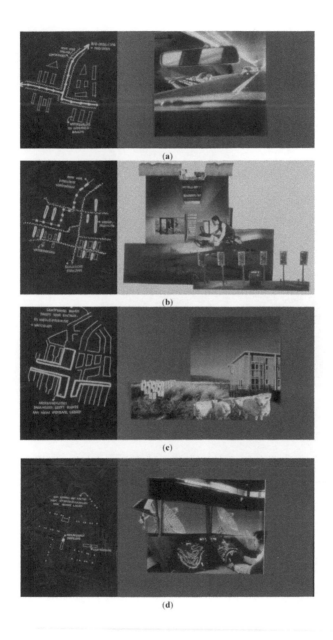

**FIGURE 4:** The four layers of the multi-layer city: (a) the underground, (b) ground level, (c) rooftop and (d) air.

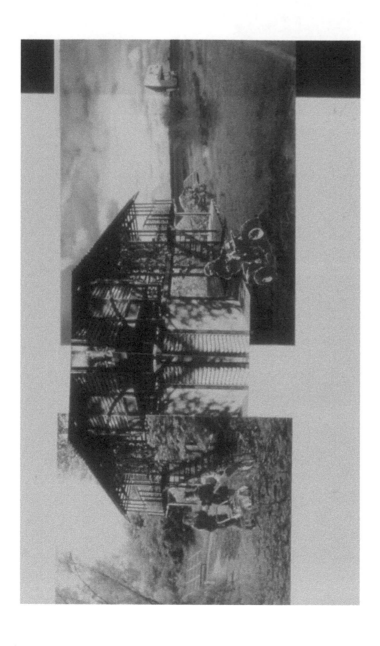

**FIGURE 5:** Light urbanism in the light city in Houten, the Netherlands.

(a)

(b)

**FIGURE 6:** The empty city [103]. (a) "The Vertical Restaurant". Dinner at the highest level. The entrées are served at ground level, coffee with cognac at the rooftop terrace. Every floor has its own course, leading to an evening-long dynamic program. With each course you encounter a new environment with plenty of rooms to choose your preferred dishes. The top is the finale, with its view over the city, splashing water and heated terrace; (b) "Out of the Row". Get three houses for the price of one! When your neighbours have left you could "use" two extra units. One and a half houses become your own and you may use the other one and a half while remaining in the ownership of the housing corporation. You are obliged to maintain the extra units and choose your desired use: a solar plant on the roof, mushrooms in the cellar, vegetables in the greenhouse on the first floor, or a handyman shop at ground level. These are all ways to earn some extra income. From a standard row of eight attached houses, the two middle units could be demolished, leaving two detached houses each the size of three original units. Instead of living in between neighbours you are suddenly surrounded by green space.

At the next level of urban environments and cities, poly-rational planning approaches provide the freedom to experiment with emergent behaviour and self-organization of urban objects. If governmental rules exclude these alternative urban developments, cities will ever face difficulties to become more resilient. Moreover, it must be physically possible to move urban objects around and allow for self-organization to reach the highest possible adaptive capacity. The city must not only be dismantable but potential configurations must also be understood and simulated (there is a need to identify the places that the objects would be likely to move to), in order to prevent random, unwanted movements.

Here, it is necessary to understand the dual complexity of the city: the city as a whole is a complex adaptive system, as is each of its parts [43,73,74]. In nature, such systems exhibit swarm behaviour [75]. Swarms [76,77] are self-organizing systems that prepare for and respond to changing circumstances, achieving this through (a) the interactions taking place between a large number of similar and free moving "agents", which (b) react autonomously and quickly to one another and their surrounding, resulting in (c) the development of a collective new entity and a coherent larger unity of higher order [75]. When swarm behaviour is applied to cities, higher levels of adaptive capacity can be achieved when active interventions on both levels of the "dually complex" city are planned. Current theory describes the process of an evolving system becoming unstable, reaching a crisis, "tipping" and transforming through self-organization to another stable state. Tipping points are identified after they have occurred [78], or identify the patterns that announce these points [79], but they are not planned. In the case of climate change, this system change would preferably be anticipated before the actual event (the disaster) occurs. Hence, an early (active) intervention would initiate self-organizing processes of the urban objects, allowing the system to "flip" to a system state with a higher adaptive capacity that is less vulnerable. Network theory (see for instance [80,81,82,83,84,85]) holds the key to identifying the intervention point [86]. The type of intervention cannot be determined other than through the local context (existing city combined with climate impact). At the level of urban objects, interactions allow them to develop emergent properties, to self-organize and to change [87]. Provided the objects are mobile, the principles of self-organization of a (urban) system will direct the city to move towards an optimal stable state, both at the level of the entire system (e.g., city) and its parts [43,73,74], as these mobile objects then always try to reach the best places in the "fitness landscape" [88,89,90]. The first attempts to use this thinking in design have been undertaken in building, attributing swarm characteristics to building components [67,68,69] and in the landscape, attributing landscape objects with complex adaptive systems properties [91]. Extended research is currently under way [92] to simulate urban precincts through agent-based modelling of mobile urban objects under the influence of an external (climate) impact that initiates the process of self-organization. Through the simulation, the ways to enhance adaptive capacity can be determined before a real hazard impacts the area.

For this to happen, it is crucial to create urban objects that can be dismantled and allowed to reconfigure when external influences require it. Therefore, physical mobility of the dismantable elements is required, and space must be created in the urban fabric to allow objects to move and nest in new locations. The next section will describe three urban concepts in which this space and mobility is created: the multi-layer, the light, and the empty city.

## 6.6 MULTI-LAYER, LIGHT AND EMPTY CITIES

Once it is understood that urban objects must have the opportunity to be mobile, existing built structures need to be adjusted in order to allow new functionalities to be developed (see for instance Section 6.3), and the urban environment needs to offer spaces where these mobile objects can move. The technical constructive research required for the former is out of scope for this article; regarding the latter, there are three urban typologies in which these new spaces are sought. While these examples may be inspiring, they also require a realization strategy to prevent them from ending in vain. The "learning by doing" strategy, in which flows, actors and areas come together, [44,93,94] may prove useful.

### 6.6.1 THE MULTI-LAYER CITY

There are many designs that propose vertical cities, wherein a single sky-scraper-type structure may contain an entire city, including gardens and vertical forests (e.g., [95]). However, very limited research is available regarding how extra space can be created in ordinary neighbourhoods for unknown future functions or urban objects seeking alternative places.

The multi-layer city is a hypothetical urban design, developed in a Design Lab environment, in which more free spaces are created by extending the useable layers. Apart from the ground level, currently used for green space, transport space, and private land, four additional layers can been identified (Figure 4). This makes it possible to move and relocate urban objects in different layers and positions when desirable. The level of choice quadruples.

The first extra layer (the underground) provides extra space for waste transportation systems, parking, water storage, energy infrastructure and the delivery of goods by car, small trucks or via a tube-system. In the second extra layer, ground-level, use is doubled, offering space to intensify residential zones, to store water, to produce food, for ecological structures or to harvest renewable energy. The third extra layer is rooftops, on which lightweight frames can be installed to harvest energy (solar, biomass) and produce food. It also offers opportunities for recreation and leisure: take a stroll on the roof, enjoying the view or play a game of tennis, as Roger Federer and Andre Agassi already showcased on the 211 m high tennis court at the Burj-al-Arab building in Dubai. The connected roof space functions as an additional public area. The final, fourth, layer is up in the air: through the use of tall poles, spaces are created where alternative objects and functions (residential units, flying-bikes, helicopters, and solar panels) can be connected. There is also space for air cleaning appliances or zeppelin-landing points.

The combination of these extra spaces offers a wide variety for objects, and people to find appropriate locations where they can move. Together they increase the resilience of the entire system.

### 6.6.2 LIGHT URBANISM

Light urbanism requires urban objects to be light, dismantable and modular, so that they can be taken up easily, moved to another place, and separated from infrastructure. This already takes place in the form of moveable or floatable houses, but not in a coherent way at an urban design scale. Elaborating the idea to "Colonize the Void" [96,97] of Randstad Holland with individual homes on large plots of gardens [98], while leaving behind the unsustainable and expensive Dutch idea to add two meters of sand for stabilizing reasons before building can commence, the planning model called "Lite Urbanism" [99,100,101,102] proposes an alternative. It consists of two typologies: "Campingland", a super-communal space that can be set up as a wood or nature reserve, and "Villageland", with large gardens around houses, enabling a village-like environment at low densities of seven to ten dwellings per hectare [99]. Both typologies make use of lighter roads and remove the majority of infrastructure, replacing telephone cables in the ground by mobile telephones, gas pipes by an electricity net connected to local alternative energy sources, and sewage plants at a distance by a water purification system in the garden. This form of light urbanism is especially useful when urban objects have a temporary location or function. Like a good traveller, a city should also be able to move without leaving a trace [54].

These principles, which allow residential units to move, have been used in the design for a light city in Houten, the Netherlands, constructed in a Design Lab. In this concept (Figure 5), autarkic light dwellings are proposed for a low-density development in a low and wet area. The light constructions and disconnection from heavy infrastructure make it possible to move houses, but also to deal with temporary wet conditions as a result of heavy rainfall. The houses are positioned on long and narrow parcels, which makes it possible to move "backwards", up onto the slightly higher parts, keeping them dry from sudden wet conditions. The light construction makes it also possible to lift the houses above temporary higher water levels.

### 6.6.3 THE EMPTY CITY

The third concept is the empty city [103]. In this example, an outdated neighbourhood, Hoge Vucht in the city of Breda, has been used to recreate viability and liveability by replacing dysfunctional elements (which create empty, vacant spaces in the city without turning it into a ghost town) by new modules and urban objects, increasing adaptability and enhancing sustainability of the entire area (Figure 6). Instead of demolishing entire neighbourhoods and replacing them by something new, this approach makes only a part of these neighbourhoods redundant and open for new urban objects or plug-ins. These new objects could be sustainable replacements, such as better, greener buildings, or new functional components producing food or energy. Sometimes physical objects are "plugged-in", but it is also possible to change functionality without changing the built structures.

This leads to increased diversity in the neighbourhood, enriching an, often monotone, environment with a range of new objects and functions. In between existing high-rise buildings, abandoned green spaces are replaced by agricultural production, which provides local food to a vertical restaurant, where dinner ends at the top floor of the same high-rise building (see Box A). The existing small flats in these high-rise buildings are rearranged into double-sized large apartments, with a huge floor space and a great view. Low-rise attached houses are deconstructed and reassembled to extend the size of the individual homes and to plug in agricultural or energy harvesting units in between (see Box B).

**Box A "The Vertical Restaurant"**
Dinner at the highest level. The entrées are served at ground level, coffee with cognac at the rooftop terrace. Every floor has its own course, leading to an evening-long dynamic program. With each course you encounter a new environment with plenty of rooms to choose your preferred dishes. The top is the finale, with its view over the city, splashing water and heated terrace.

**Box B "Out of the Row"**
Get three houses for the price of one! When your neighbours have left you could 'use' two extra units. One and a half houses become your own; the other one and a half may be used by you while remaining in the ownership of the housing corporation. You are obliged to maintain the extra units and choose your desired use: a solar plant on the roof, mushrooms in the cellar, vegetables in the greenhouse on the first floor, or a handyman shop at ground level. These are all ways to earn some extra income. From a standard row of eight attached houses, the two middle units could be demol-

ished, leaving two detached houses each the size of three original units. Instead of living in between neighbours you are suddenly surrounded by green space.

These transformations, despite their aim of increasing sustainability, can be quite confronting for local citizens. In this design project, people were therefore invited to literally stand model, by having their picture taken, for a specific transformation (Figure 6). Every specific spatial modification proposed in the neighbourhood was printed on foam boards was explained to citizens, before they were taken on picture, created around 25 "change stories" increased understanding and support for the design project.

## 6.7 CURRENT CONSTRAINTS

The proposal of a Dismantable City is a new concept and it is obvious that it cannot be implemented immediately due to several considerable constraints.

First, the current technological conditions in cities are not conducive to dismantability. Many of the city elements are difficult to loosen. Most buildings are fixed to the underground or next-door neighbours. In addition, the majority of the infrastructure is inextricably connected to its environment. Current practice in building infrastructure and buildings aims to strengthen constructions, which makes them inevitably more strongly connected to their surroundings. These construction "habits" are not easily changed. To speculate on how to get around this constraint is allowing building of moveable objects free of the land-use plan or building codes, e.g., when a developer, builder or contractor proposes a construction that can easily move to another place, it doesn't have the restrictions another, more fixed, object has. The building code must be abandoned for moveable objects.

Secondly, there are social constraints. Only a limited number of people are willing to live in a place that will potentially move. Despite the romance of living on a houseboat or a ship, in reality not many people choose a floating residence, principally because few are familiar with the potential safety benefits. Moreover, these types of houses are currently not available in large numbers. In practice, only very specific and small projects that are owned and developed by the initiators are being developed. This means that individuals need to put a lot of effort into developing a project themselves. At the same time, ordinary fixed houses are available in large numbers, making them a more accessible option for an average person. A way to mitigate this effect is to offer spectacular zones, such as on the beach, dune tops, hilltops and more places that are vulnerable but with great views, to houses that are easy to move and relocate.

The third constraint is found in the current regulatory frameworks, which prevent the development of moveable urban elements. In the majority of cases, land use is regulated, determining what uses are allowed and, more importantly, prohibited. Changing the permissible land use requires revision of the plan, which is generally not possible within the term of the plan. Quick alterations in use are not supported, which makes it difficult to allow short-term change and/or reverse the change soon after. It is even more difficult to regulate multiple uses for a given area, alternating between two or more functions. To speculate a bit more, the legal framework must be adjusted to create the possibility of double land-use and the permission of temporary use.

These constraints are current and probably not easy to abrogate, but must be considered in order to create the space to respond to unprecedented occurrences. As long as these constraints imply creating an inflexible city, the city cannot respond, or even anticipate a future that is increasingly uncertain.

## 6.8 CONCLUSION

The Dismantable City is an innovative concept for building more resilient cities. It marks a fundamental shift in urban design. From the comprehensive and predictable planning practice in the technical-rational era, the Dismantable City absorbs interferences and provides flexibility. This break with history is necessary because current environmental issues are severe and demand new concepts. Extreme events, of which we expect many in the future, simply play around with our existing cities. The way our current cities are built is not only unsustainable, with a high metabolism, but also lacks the flexibility to be adjusted when necessary. This article has only given a first glimpse of the potential to rebuild our cities in a more resilient way and it is obvious that more work needs to be done in order to make this proposition more realistic. However, it is also obvious that continuation of current practices only leads to repetition in urban design and implies that we keep on rebuilding our cities after disasters. Fundamental rethinking of the way our cities are built is therefore essential and will create less vulnerable communities, and also neighbourhoods that are more interesting to live in, socially more inclusive and ecologically more productive.

Some questions remain to be answered in future research and experiments:

1. Further design experiments are necessary to discover the conditions for dismantable cities in practical applications and to explore the potential to build realistic examples.

2. The urban fabric, being part of a flexible and changeable environment is, in itself, subject to change. Elaboration is required to discover what ultra-high adaptability of urban environments might comprehend. Dynamic modelling might shine a light on eventual general rules and likely configurations.

3. The urban metabolism model as it has been approached in this article, assumes that increased mobility of urban elements leads to a greater flexibility of and in cities, allowing for a more efficient metabolism, e.g., lesser and better quality of resources required and leading to less waste, and better liveability (for example: safety) of the city, which implies a more sustainable city. However, further research is required to estimate metabolisms of pilot studies in order to underpin this assumption.

4. Despite research is not yet available on the metabolism of the Dismantable City, a comparison with a "rebuildable" city, e.g., the energy and materials it takes to rebuild a city after a disaster has destroyed the city, could deliver insights in the assumed benefits for the metabolism of the Dismantable City.

5.  The main future challenge for engineering consultants and companies is to overcome and further research the technical constraints of creating more mobile urban objects.

The Dismantable City faces challenges, but given that the majority of the current global population lives in urban environments and faces the impact of climate change, realisation of highly resilient urban environments in the form of Dismantable Cities could represent a very attractive future for many. Especially when cities grow fast, the mobility of urban objects could prove useful, as detached buildings and urban objects are easier, lighter, faster and cheaper to build. These objects allow a fast building pace and, as these rapid urban developments are often located in vulnerable zones for the impacts of climate change, offer the chance to relocate the city, or neighbourhoods, more easily. Moreover, moveable objects are less susceptible to hazards as they are lighter and could even anticipate a hazard by simply moving away.

## REFERENCES

1.  Walker, B.; Holling, C.S.; Carpenter, S.R.; Kinzig, A. Resilience, adaptability and transformability in social-economic systems. Ecol. Soc. 2004, 9, p. 5. Available online: http://www.ecologyandsociety.org/vol9/iss2/art5/ (accessed on 21 February 2010).
2.  Walker, B.; Salt, D. Resilience Thinking; Island Press: Washington, DC, USA, 2006.
3.  Post Carbon Institute. Resilience. Available online: www.resilience.org (accessed on 2 March 2012).
4.  Carver, C.S. Resilience and thriving: Issues, models and linkages. J. Soc. Issues 1998, 54, 245–266.
5.  Tugade, M.M.; Fredrickson, B.L.; Feldman Barrett, L. Psychological resilience and positive emotional granularity: Examining the benefits of positive emotions on coping and health. J. Personal. 2004, 72, 1161–1190.
6.  Mubeen, A. Mechanics of Solids, 2nd ed.; Pearson Education India: Delhi, India, 2011.
7.  Hollnagel, E., Woods, D.D., Leveson, N., Eds.; Resilience Engineering: Concepts and Precepts; Ashgate Publishing Ltd: Aldershot, ON, Canada, 2006.
8.  Newman, P.W.; Birrell, R.; Homes, D.; Mathers, C.; Newton, P.; Oakley, G.; O'Connor, A.; Walker, B.; Spessa, A.; Tait, D. Human settlements. In Australia: State of the Environment 1996; Taylor, R., Ed.; Department of Environment, Australian Government: Canberra, ACT, Australia, 1996; Chapter 3. pp. 1–57.

9.   Newman, P.; Beatley, T.; Boyer, H. Resilient Cities, Responding to Peak Oil and Climate Change; Island Press: Washington, DC, USA, 2009.

10.  Holling, C.S. Understanding the complexity of economic, ecological and social systems. Ecosystems 2001, 4, 390–405.

11.  The Resilience Alliance. Adaptive Cycle. Available online: www.resalliance.org/index.php/adaptive_cycle (accessed on 2 March 2012).

12.  Department for Environment, Food & Rural Affairs (DEFRA). UK Climate Change Risk Assessment 2012; The Stationery Office Limited: London, UK, 2012.

13.  KPMG. Australia Report: Risk Landscape. Available online: www.kpmg.com/au/en/issuesandinsights/articlespublications/australia-report/pages/risk-landscape-2012.aspx (accessed on 12 October 2013).

14.  Roggema, R. Swarm Planning: The Development of a New Planning Methodology to Deal with Climate Adaptation. Ph.D. Thesis, Delft University of Technology and Wageningen University and Research Center, Delft, the Netherlands, 2012.

15.  Biggs, C.; Ryan, C.; Bird, J.; Trudgeon, M.; Roggema, R. Visions of Resilience: Design-Led Transformation for Climate Extremes; NDR-Final Report. The University of Melbourne: Melbourne, VIC, Australia, 2014.

16.  Comfortable Chaos. Available online: http://comfortablechaos.tumblr.com (accessed on 3 May 2013).

17.  Prigogine, I. Civilisation and democracy: Values, systems, structures and affinities. Futures 1986, 18, 493–507.

18.  Prigogine, I.; Stengers, I. Order out of Chaos, 1st ed.; Bantam Books: New York, NY, USA, 1984.

19.  Kauffman, S. The Origins of Order; Oxford University Press: New York, NY, USA, 1993.

20.  Epstein, J.M.; Axtell, R.L. Growing Artificial Societies: Social Science from the Bottom up; Brookings Institution: Washington, DC, USA, 1996.

21.  Lansing, J.S. Complex adaptive systems. Annu. Rev. Anthropol. 2003, 32, 183–204.

22.  Liu, J.; Dietz, T.; Carpenter, S.R.; Alberti, M.; Folke, C.; Moran, E.; Pell, A.N.; Deadman, P.; Kratz, T.; Lubchenco, J.; et al. Complexity of coupled human and natural systems. Science 2007, 316, 1513–1516.

23.  Nowotny, H. The increase of complexity and its reduction: Emergent interfaces between the natural sciences, humanities and social sciences. Theory Cult. Soc. 2005, 22, 15–31.

24.  Urry, J. Complexity. Theory Cult. Soc. 2006, 23, 111–115.

25.  Anderson, P. Complexity theory and organisation science. Organ. Sci. 1999, 10, 216–232.

26.  Byrne, D. Complexity theory and planning theory: A necessary encounter. Plan. Theory 2003, 2, 171–178.

27.  Crawford, T.W.; Messina, J.P.; Manson, S.M.; O'Sullivan, D. Guest editorial. Environ. Plan. B 2005, 32, 792–798.

28.  Duit, A.; Galaz, V. Governance and complexity—Emerging issues for governance theory. Governance 2008, 21, 311–335.

29.  Levinthal, D.A.; Warglien, M. Landscape design: Designing for local action in complex worlds. Organ. Sci. 1999, 10, 342–357.

30. Montalvo, C. What triggers change and innovation? Technovation 2006, 26, 312–323.
31. O'Sullivan, D. Complexity science and human geography. Trans. Inst. Br. Geogr. NS 2004, 29, 282–295.
32. Plowman, D.A.; Baker, L.T.; Beck, T.E.; Kulkarni, M.; Solansky, S.T.; Travis, D.V. Radical change accidentally: The emergence and amplification of small change. Acad. Manag. J. 2007, 80, 515–543.
33. Pulselli, R.M.; Tiezzi, E. City Out of Chaos; WIT Press: Southampton, UK, 2009.
34. Richards, A. Complexity in physical geography. Geography 2002, 87, 99–107.
35. Teisman, G.R.; Klijn, E.H. Complexity theory and public management. Public Manag. Rev. 2008, 10, 287–297.
36. Timmermans, W.; van Dijk, T.; van der Jagt, P.; Onega Lopez, F.; Crecente, R. The unexpected course of institutional innovation processes: Inquiry into innovation processes into land development practices across Europe. Int. J. Des. Nat. Ecodynamics 2011, 6, 297–317.
37. Timmermans, W.; Ónega López, F.; Roggema, R. Complexity theory, spatial planning and adaptation to climate change. In Swarming Landscapes: The Art of Designing for Climate Adaptation; Roggema, R., Ed.; Springer: Heidelberg, Germany, 2012; pp. 43–65.
38. de Jonge, J. Landscape Architecture between Politics and Science, an Integrative Perspective on Landscape Planning and Design in the Network Society. Ph.D. Thesis, Wageningen University, Amsterdam, the Netherlands, 2009.
39. de Roo, G.; Porter, G. Fuzzy Planning, the Role of Actors in a Fuzzy Governance Environment; Ashgate Publishing: Burlington, VT, Canada, 2007.
40. Roggema, R.; van den Dobbelsteen, A. Swarm planning: Development of a new planning paradigm, which improves the capacity of regional spatial systems to adapt to climate change. In Proceedings of the World Sustainable Building Conference (SB08), Melbourne, VIC, Australia, 21–25 September 2008.
41. Timmermans, W. Crisis and innovation in sustainable urban planning. In Advances in Architecture; WIT-Press: Southampton, UK, 2004; Volume 18, pp. 53–63.
42. Odum, H.T. Systems Ecology: An Introduction; John Wiley and Sons: New York, NY, USA, 1983.
43. Portugali, J. Self-Organisation and the City; Springer-Verlag: Berlin/Heidelberg, Germany, 2000.
44. Tjallingii, S.P. Ecopolis: Strategies for Ecologically Sound Urban Development; Backhuys Publishers: Leiden, the Netherlands, 1993.
45. Wolman, A. The metabolism of cities. Sci. Am. 1965, 213, 178–193.
46. Newman, P.W.G. Sustainability and cities: Extending the metabolism model. Landsc. Urban Plan. 1999, 44, 219–226.
47. Minx, J.; Creutzig, F.; Medinger, V.; Ziegler, T.; Owen, A.; Baiocchi, G. Developing a Pragmatic Approach to Assess Urban Metabolism in Europe; A Report to the European Environment Agency. 2010. 2010. 2010.
48. Newman, P. The environmental impact of cities. Environ. Urban. 2006, 18, 275–295.
49. Newton, P.W., Ed.; Re-Shaping Cities for a More Sustainable Future; Research Monograph 6. Australian Housing and Urban Research Institute (AHURI): Melbourne, VIC, Australia, 1997.

50. Newton, P.; Flood, J.; Berry, M.; Bhatia, K.; Brown, S.; Cabelli, A.; Gomboso, J.; Higgins, J.; Richardson, T.; Ritchie, V. Environmental indicators for national state of the environment reporting—Human settlements. In Australia: State of the Environment (Environmental Indicator Reports); Department of Environment, Australian Government: Canberra, ACT, Australia, 1998; pp. 1–188.

51. Newton, P.W.; Baum, S.; Bhatia, K.; Brown, S.K.; Cameron, A.S.; Foran, B.; Grant, T.; Mak, S.L.; Memmott, P.; Mitchell, V.C.; Neate, K.; et al. Human settlements. In Australia State of the Environment Report 2001 (Theme Report); CSIRO Publishing on Behalf of the Department of the Environment, Australian Government: Canberra, ACT, Australia, 2001; pp. 1–198.

52. Niza, S.; Rosado, L.; Ferrão, P. Urban metabolism. J. Ind. Ecol. 2009, 13, 384–405.

53. Tomásek, W. Die Stadt als Oekosystem; Überlegungen zum Vorentwurf Landschafsplan Köln (The city as ecosystem; considerations about the scheme of the Landscape design Cologne). Landsch. Stadt 1979, 11, 51–60.

54. Kristinsson, J. Integrated Sustainable Design; van den Dobbelsteen, A., Ed.; Delft Digital Press: Delft, the Netherlands, 2012.

55. Merry, U. Coping with Uncertainty: Insights from the New Sciences of Chaos, Self-Organisation and Complexity; Praeger: Westport, CT, USA, 1995.

56. Davy, B. Plan it without a condom! Plan. Theory 2008, 7, 301–317.

57. Gunder, M. Fake it until you make it, and then…. Plan. Theory 2011, 10, 201–212.

58. Newman, S. Post-anarchism and space: Revolutionary fantasies and autonomous zones. Plan. Theory 2011, 10, 344–365.

59. Boelens, L. Theorizing practice and practising theory: Outlines for an actor-relational-approach in planning. Plan. Theory 2010, 9, 28–62.

60. Miraftab, F. Insurgent planning: Situating radical planning in the Global South. Plan. Theory 2009, 8, 32–50.

61. Timmermans, W.; Cilliers, J.; Slijkhuis, J. Planning by Surprise: The Values of Green Spaces in Towns and Cities; Van Hall Larenstein: Velp, the Netherlands, 2012.

62. Cook, P.; Webb, M. Archigram; Princeton Architectural Press: Princeton, NJ, USA, 1999.

63. Sadler, S. Archigram: Architecture without Architecture; MIT Press: Cambridge, UK, 2005.

64. Superuse Studios. Available online: www.superuse-studios.com (accessed on 4 May 2013).

65. Uhde, R. Secondhand architektur. Aface 2012, 6, 35–36.

66. Leroux, K.; Seldman, N. Deconstruction, Salvaging Yesterday's Buildings for Tomorrow's Sustainable Communities; Institute for Local Self-Reliance: Washington, DC, USA, 1999.

67. Oosterhuis, K. Swarm architecture. In Game, Set and Match II, On Computer Games, Advanced Geometries and Digital Technologies (No.2); Oosterhuis, K., Feireiss, L., Themans, M., Eds.; Episode Publishers: Rotterdam, the Netherlands, 2006; Chapter II. pp. 14–28.

68. Oosterhuis, K. Towards a New Kind of Building, A Designer's Guide to Nonstandard Architecture; NAI Uitgevers: Rotterdam, the Netherlands, 2011.

69. Oosterhuis, K. Hyperbody: First Decade of Interactive Architecture; Ram Publications: Santa Monica, CA, USA, 2012.

70. Wiki-Habitat 67. Available online: http://en.wikipedia.org/wiki/Habitat_67 (accessed on 7 May 2013).
71. JafUd.org. Available online: www.jafud.org (accessed on 27 December 2006).
72. Roggema, R. Swarming landscapes. In Swarming Landscapes: The Art of Designing for Climate Adaptation; Roggema, R., Ed.; Springer: Heidelberg, Germany, 2012; pp. 167–193.
73. Portugali, J. Complexity theory as a link between space and place. Environ. Plan. A 2006, 38, 647–664.
74. Portugali, J. Learning from paradoxes about prediction and planning in self-organising cities. Plan. Theory 2008, 7, 248–262.
75. van Ginneken, J. De kracht van de Zwerm (The Power of the Swarm); Antwerpen: Amsterdam, the Netherlands, 2009.
76. Fisher, L. The Perfect Swarm, the Science of Complexity in Everyday Life; Basic Books: New York, NY, USA, 2009.
77. Miller, P. The Smart Swarm; The Penguin Group: New York, NY, USA, 2010.
78. Gladwell, M. The Tipping Point; Little, Brown and Company, Time Warner Book Group: New York, NY, USA, 2000.
79. Scheffer, M. Critical Transitions in Nature and Society; Princeton University Press: Princeton, NJ, USA, 2009.
80. Barabási, A.-L. Linked: How Everything is Connected to Everything Else and What it Means for Business, Science, and Everyday Life; Penguin Group: London, UK, 2002.
81. Bianconi, G.; Barabási, A.-L. Competition and multiscaling in evolving networks. Europhys. Lett. 2001, 54, 436–442.
82. Broder, A.; Kumar, R.; Maghoul, F.; Raghavan, P.; Rajagopalan, S.; Stata, R.; Tomkins, A.; Wiener, J. Graph structure in the Web. Comput. Netw. 2000, 33, 309–320.
83. Erdós, P.; Rényi, A. On the evolution of random graphs. Publ. Math. Inst. Hung. Acad. Sci. 1960, 5, 17–61.
84. Solé, R.V.; Pastor-Satorras, R.; Smith, E.; Kepler, T.B. A model of large-scale proteome evolution. Adv. Complex Syst. 2002, 5, 43–54.
85. Watts, D.J.; Strogatz, S.H. Collective dynamics of "small-world" networks. Nature 1998, 393, 440–442.
86. Roggema, R.; Vermeend, T.; van den Dobbelsteen, A. Incremental change, transition or transformation? Optimising change pathways for climate adaptation in spatial planning. Sustainability 2012, 4, 2525–2549.
87. Manson, S.M. Simplifying complexity: A review of complexity theory. Geoforum 2001, 32, 405–414.
88. Cohen, J.; Stewart, I. The Collapse of Chaos: Discovering Simplicity in a Complex World; Penguin Books: London, UK, 1995.
89. Langton, C.G.; Taylor, C.; Farmer, J.D.; Rasmussen, S. Artificial Life II (Studies in the Sciences of Complexity Proceedings); Santa Fe Institute: Redwood City, CA, USA, 1992; Volume 10.
90. Mitchell Waldrop, M. Complexity. The Emerging Science at the Edge of Order and Chaos, 1st ed.; Simon & Schuster: New York, NY, USA, 1992.

91. Roggema, R. Swarming landscapes, new pathways for resilient cities. In Proceedings of the 4th International Urban Design Conference—Resilience in Urban Design, Surfers Paradise, Gold Coast, QLD, Australia, 23 September 2011.

92. Roggema, R. Modelling the Benefits of Swarm Planning, a Dynamic Way of Planning for Future Adaptation to Climate Change Impacts. Rubicon Research Proposal (Granted); NWO: Den Haag, the Netherlands, 2012.

93. Tjallingii, S.P. Carrying structures: Urban development guided by water and traffic networks. In Shifting Sense, Looking Back to the Future in Spatial Planning; Hulsbergen, E.D., Klaasen, L.T., Kriens, L., Eds.; Techne Press: Amsterdam, the Netherlands, 2005; pp. 355–368.

94. Tjallingii, S.P.; de Roo, G. Complexity and carrying structures—Learning-by-doing in urban planning. In Presented at AESOP Working Group, Stockholm, Sweden, 26–27 February 2010.

95. Aiello, C.; Aldridge, P.; Deville, N.; Solt, A.; Lee, J.S. Evolo Skyscrapers; Evolo Inc.: Los Angeles, CA, USA, 2012; Volumes 1 and 2.

96. Geuze, A. Wildernis (Wilderness). In Alexanderpolder, New Urban Frontiers; Devolder, A.-M., Ed.; Uitgeverij THOTH: Bussum, the Netherlands, 1993; pp. 96–105.

97. van Dijk, H.; Geuze, A.; Bindels, E.; Musch, J. Colonizing the Void; NAI Publishers: Rotterdam, the Netherlands, 1996.

98. Lootsma, B. Synthetic regionalisation: The Dutch landscape toward a second modernity. In Recovering Landscape, Essays in Contemporary Landscape Architecture, 1st ed.; Cornor, J., Ed.; Princeton Architectural Press: New York, NY, USA, 1999; pp. 250–274.

99. Maas, W. Light urbanism. Archis 1997, 11, 74–79.

100. Moreno Mansilla, L.; Tuñon, E. Una conversación con Winy Maas, Jacob van Rijs y Nathalie de Vries (A conversation with Winy Maas, Jacob van Rijs and Nathalie de Vries). In MVRDV; Levene, R.C., Márquez Cecilia, F., Eds.; El Croquis: Madrid, Spain, 1997; Volume 58, pp. 6–25.

101. Maas, W.; van Rijs, J.; Koek, R. Permanence. In Farmax: Excursions on Density; 010 Publishers: Rotterdam, the Netherlands, 1998; pp. 34–51.

102. MVRDV. Architects & Kristinsson Engineers. Midden-IJsselmonde, Lichte Stedebouw. Smitshoek en de toekomst van de Nederlandse uitbreidingswijk (Midden-IJsselmonde, Light Urban Planning. Smitshoek and the future of the Dutch new development district). In Rotterdam 2045, Visies op de Toekomst van Stad, Haven en Region (Rotterdam 2045, Visions of the Future for the Town, Harbour and Region); Boekraad, C., van Es, W., Eds.; Netherlands Architecture Institute (NAI): Rotterdam, the Netherlands, 1995.

103. Roggema, R.; Timmermans, W.; Oost, L. Transformatie Hoge Vught. (Transformation of Hoge Vught); IBN Rapport 450. IBN-DLO: Wageningen, the Netherlands, 1999.

# CHAPTER 7

# Considering Hazard Estimation Uncertain in Urban Resilience Strategies

B. BARROCA, P. BERNARDARA, S. GIRARD, AND G. MAZO

## 7.1 INTRODUCTION

Climate change, combined with a higher concentration of property and persons in urban areas and the increasing sensitiveness of our urban systems, foretell devastating events for the years to come. By the end of the century, the economic cost of flood risks throughout the world is liable to attain a value of EUR 100 billion per year (EEA, 2011).

Aside from exceptional cases, de-urbanizing flood areas is out of the question due to economic development (Klein et al., 2004), social acceptance (Adger et al., 2008) and the environmental challenges raised by sustainable development, which include limiting urban sprawl by increasing city density and compactness. Therefore, the fight against damage caused by flooding, as well as the sustainable development objectives that apply to urban technical systems, mean that resilience actions must be implemented (Milman and Short, 2008). If hazards prove to be interesting factors of innovation for cities and buildings (Romero-Lankao and Dodman, 2011), risk management measures must be taken in an appropriate context of governance and with adequate knowledge of any changes in socio-economic contexts and uncertainties (Adger et al., 2008). Research on vulnerability has increased over the last few years (Serre and Barroca, 2013; Birkmann et al., 2013). This type of research normally assesses a city's vulnerability to a hazard and sometimes introduce resilience indicators, strategies or adaptation scenarios (Barroca et al., 2006; Romero-Lankao and Qin, 2011). If various authors agree to admit that, for anticipating flooding efficiently, implementation of resilient strategies must anticipate flooding scenarios, which today's probabilistic models deem to be extreme or rare (Zevenbergen et al., 2011), it would appear necessary to put the reliability of these results into question.

For modelling hazards, especially hydrological hazards, we cannot exclude important uncertainties, especially when modelling rare events (Barroca, 2006). Improving risk management for events that possess considerable evaluation uncertainty must integrate this uncertainty into strategic orientations. In this article, strategic analysis is developed by characterizing regions for implementing resilience by incorporating uncertainty in hazard evaluations. This article does not deal with the holistic problem of resilience, which involves cultural, social, environmental, economic and institutional resilience and the link between the various facets. To implement a local strategy, the central aim of this article is to develop an approach for understanding the importance of urban components and critical infrastructures.

A method for evaluating the uncertainty due to extreme events is presented in Sect. 2 and is illustrated in Sect. 3 on the Besançon data set. A guiding action for regional resilience is proposed in Sect. 4 and concluding remarks are provided in Sect. 5.

## 7.2 STATISTICAL EVALUATION OF UNCERTAINTY

On several rivers, the high discharges observed over recent years exceed the prediction of very rare quantile carried out in the past by hydrologists. Two main explanations exist:

- The floods are very extreme and their probability of occurrence is very small.
- These floods are important, but their "beyond the norm" nature is merely wishful thinking. This illusion is perpetuated by errors inherent to estimates of their return period which, on the face of it, are too great.

To identify a flood-prone area in the event of a rise in water levels—100-year flooding for example—we need to make a series of analyses and choices. Uncertainties exist at every stage, which makes estimating global uncertainty an extremely complex task. This section presents the characterization of uncertainties, especially the uncertainty as to the choice of mathematical model to be used for estimating the hazard.

We will not go into measurement uncertainty (Lang et al., 2006; Gaume et al., 2004) nor the validity of sometimes obsolete measurements in a context of climate change. Uncertainties on the physical model are generally circumscribed, but uncertainties related to the choice of mathematical model used for estimating extreme flow rates are not presented in risk analyses. Hydrologists' culture (in the sense of their usual habits) leads them to systematically use the so-called Gumbel model without assessing its relevance in the face of data distribution (Payrastre et al., 2005; Payrastre, 2005; Bernardara et al., 2008).

## 7.2.1 EXTREME-VALUE THEORY

Extreme-value theory is a relevant tool for estimating N-year return level (denoted by TN ) of floods or rainfalls when N is larger than the number of years of observations. In such a case, TN is beyond the observation range and extrapolation is thus needed. Extreme-value theory provides several estimators as well as evaluations of their associated uncertainty through the construction of confidence intervals. Two types of methods are available; see Coles (2001) for further details.

### 7.2.1.1 BLOCK MAXIMA APPROACH

Let $X_1$, $X_2$, ... be a sequence of independent random variables with a common distribution function F. Denoted by $M_n = \max(X_1,...,X_n)$ their maxima with distribution function $F^n$. The extreme-value theory states that the distribution function of their maxima can be approximated by the generalized extreme value (GEV) distribution function defined as

$$G_\xi = \exp\left[-\left(1 + \xi\left(\frac{x - \mu}{\sigma}\right)\right)^{-1/\xi}\right]$$

(1)

for all x such that $1 + \xi(x - \mu)/\sigma > 0$. Here, $\mu$ is the location parameter, $\sigma > 0$ is the scale parameter and $\xi$ is the shape parameter referred to as the extreme-value index. In the particular case where $\xi = 0$, the GEV reduces to a Gumbel distribution:

$$G_0(x) = \exp\left[-\exp\left(-\left(\frac{x - \mu}{\sigma}\right)\right)\right]$$

(2)

Otherwise, the GEV distribution is called a Fréchet distribution ($\xi > 0$) or a Weibull distribution ($\xi < 0$). In practice, the original data $X_1$, $X_2$,... are split into m blocks of size n. For instance, a block may correspond to a time period of length one year. In such a case, n is the number of observations per year and thus the block maxima are annual maxima.

The N-year return period is then obtained by inverting $G_\xi$ at point 1/N:

$$T_N = \mu - \frac{\sigma}{\xi}\left[1 - \left(-\log\left(1 - \frac{1}{N}\right)\right)^{-\xi}\right] \simeq \mu - \frac{\sigma}{\xi}\left[1 - N^\xi\right] \quad \text{if } \xi \neq 0$$

(3)

or

$$T_N = \mu - \sigma\log\left(-\log\left(1 - \frac{1}{N}\right)\right) \simeq \mu + \sigma\log N \quad \text{if } \xi \neq 0$$

(4)

the previous approximations being reliable if N is large.

In practice, the parameters ($\mu$, $\sigma$, $\xi$) have to be estimated. Several techniques exist, the two most popular being maximum likelihood and probability weighted moments. Both of them require an interactive procedure to compute the estimators. In each case, confidence intervals on return period to assess the statistical uncertainty of the estimation. However, in the block maxima approach, the estimation depends on the choices made by the user: the size of the blocks, the assumption made on the extreme-value index ($\xi \neq 0$ or $\xi = 0$) and the estimator used (maximum likelihood or probability weighted moments). Unfortunately, there is no mathematical tool to assess the uncertainty related to these choices.

## 7.2.1.2 PEAKS OVER THRESHOLD (POT) APPROACH

The previous block maxima approach relies on the modelling of one single observation in each block: the maxima. There might be a loss of information if more than one observation is extreme in a block. To overcome this limitation, the POT approach relies on the modelling of the excesses over a threshold u. More specifically, the distribution of the $Y_i = X_i - u$ given $X_i > 0$ can be approximated by a generalized Pareto distribution (GPD) with distribution function given by

$$H_\xi(x) = 1 - \left(1 + \xi\frac{x}{\lambda}\right)^{-1/\xi}$$

(5)

for all $x > 0$, such that $1 + \xi\, x/\lambda > 0$. Here $\lambda > 0$ is a scale parameter which can be expressed as a function of the GEV parameters as $\lambda = \sigma + \xi(u-\mu)$. The shape parameters coincides with the one of the GEV distribution. In the particular case where $\xi = 0$, the GPD reduces to an exponential distribution:

$$H_0(x) = 1 - \exp(-x/\lambda)$$

(6)

Letting $p = P(X > u)$ and recalling that, for $x > u$,

$$H_0(x - u) \simeq P(X_i > x | X_i > u) = P(X_i > x)/P(X_i > u)$$

(7)

one gets the approximation $P(X_i > x) \simeq pH_\xi(x - u)$. The N-year return period can then be obtained by inverting this formula at point $1/N$:

$$T_N = u - \frac{\lambda}{\xi}\left[1 - (Np)^\xi\right] \quad \text{if } \xi \neq 0$$

(8)

or

$$T_N = u + \lambda \log(Np) \quad \text{if } \xi = 0$$

(9)

In practice, the proportion p of observations exceeding the threshold u is fixed by the user. Then, the threshold is estimated by the corresponding empirical quantile. The two remaining parameters ($\lambda$, $\xi$) are estimated as previously via maximum likelihood or probability weighted moments. Here, the estimators are closed form, their computation is straightforward. Similarly to the block maxima approach, it is possible to compute confidence intervals on return period to assess the statistical uncertainty of the estimation. Again, the estimation depends on the choices made by the user: the proportion p of excesses, the assumption made on the extreme-value index ($\xi \neq 0$ or $\xi = 0$) and the estimator used (maximum likelihood or probability weighted moments).

## 7.3 DISCUSSION

As a conclusion, extreme-value theory offers a nice framework for the estimation of return levels $T_N$ via block maxima or excesses modelling.

The expressions of $T_N$ are similar for the two approaches: u corresponds to $\mu$ while $\lambda$ corresponds to $\sigma$ in the block maxima technique. It appears that the only difference between both methods relies on the estimation of the parameters. The POT approach benefits from easy implementation due to the existence of closed-form estimators. Extreme-value theory also permits a partial evaluation of the statistical uncertainty. However, the uncertainty may be under-estimated since the variability induced by the many choices left to the user is not taken into account. The variability can also be reduced by taking into account some covariate information such as geographical location (see Ceresetti et al., 2012; Gardes and Girard, 2010).

## 7.4 APPLICATION TO THE BESANÇON ANALYSIS

Besançon is a very important town, established during the Gallo-Roman period (with the name of Vesontio) and located in eastern France in a unique geographical location. In the centre is a meander of the Doubs River, which is almost a kilometre in diameter in the shape of an almost perfect closed loop virtually forming a peninsular and dominated by Mount Saint-Étienne, a high plateau facing the Jura mountains. At present, Besançon is the 30th largest city of France with 117 392 inhabitants. It is considerably prone to flooding. The Doubs River 1910 flood, which occurred on 20 and 21 January of that year in the heart of the Franche-Comté region, is the reference used today. The 1910 water levels flooded half the city to levels of up to 1.5 m deep, or 72 cm higher than the previous 1882 floods. Historical research reveals that important floods also occurred in 1364, 1456, 1570, 1776, 1789 and 1802.

Besançon possesses a flood risk prevention plan for adapting its risk management policy.

## 7.5 DATA DESCRIPTION

Over 75 years of data on flow rates have been used for developing the flood risk prevention plan in Besançon. The Doubs River 100-year flood at Besançon is estimated as having a flow rate of 1750 $m^3$ per second, whereas the flow rate for the 1910 flood, the most serious flood known, was estimated at 1610 m³ per second. Soil sealing as a result of urbanization increases run-off and restricts infiltration. Heavy rainfall results in so-called storm water flooding locally and generally an increase in downstream water flow that can induce so-called river floods caused by overflowing.

The data used for estimating extreme flow rates using the method presented above come from the hydrological data bank which is the reference flow-rate database in France.

## 7.5.1 NUMERICAL ILLUSTRATION—RESULTS

In our block maxima implementation, each block corresponds to one year. We thus have 59 maxima to fit the GEV distribution. The estimation of the return levels is displayed in Fig. 1. On the top panel, the sample estimations (crosses) are compared to the estimation with the G$\xi$ GEV model (continuous line). On the bottom panel, they are compared to the G0 Gumbel model. It appears that, in both cases, sample estimates and model estimates are very close. Moreover, the sample estimates are always included in the corresponding 95 % confidence intervals. These results indicate a very good fit of the GEV model for $\xi \neq 0$ or $\xi = 0$. The estimated 100- year return levels are reported in Table 1. They are compared to these obtained with the probability weighted moments estimation of the GEV parameters.

**TABLE 1:** Estimated 100-year return levels.

|  | Block maxima | Peaks other threshold |
|---|---|---|
| Maximum likelihood |  |  |
| $\xi \neq 0$ | 1331 | 1341 |
| $\xi = 0$ | 1209 | 1525 |
| Probability weighted moments | 1235 | 1328 |

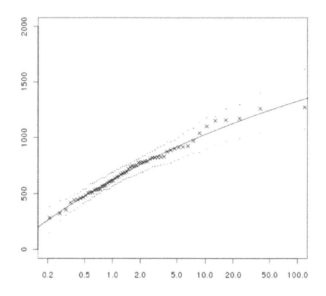

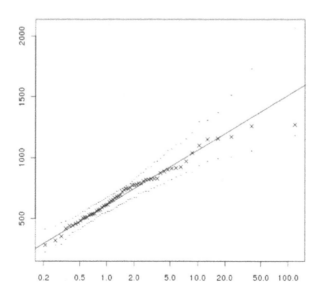

**FIGURE 1:** Return level plot (block maxima approach, maximum likelihood estimators). Top: assumption $\xi \neq 0$, bottom: assumption $\xi = 0$. Horizontally: $-1/\log(1-1/N)$, logarithmic scale, vertically: estimated N-year return level. Continuous line: estimation using the GEV, crosses: sample points, dots: 95 % confidence interval.

Turning to the POT approach, the first step is the selection of an appropriate threshold u. The selection is achieved using the mean excess function defined as $m(t) = E(X−t|X > t)$. It is known that this function should be linear for all $t > u$. The method consists in plotting an estimation of $m(t)$ and choosing u as the smallest value for which $m(t)$ is linear for all $t > u$. The graph of the so-called mean residual life plot is depicted in Fig. 3. Taking the confidence intervals into account, it appears the graph curves between $t = 0$ and $t = 350$. Beyond this interval, the graph is approximately linear until $t = 900$. However, the estimation is very unstable for $t > 900$, since it is based on very few points. This well-known phenomena is confirmed by the wide confidence intervals. We thus choose to work with a threshold fixed at $u = 350$ leading to 867 excesses. We refer to Neves and Fraga Alves (2004) for a discussion on automatic methods for selecting the threshold. The estimation of the return levels with the corresponding GPD approach is displayed in Fig. 2. On the top panel, the sample estimations (crosses) are compared to the estimation with the H$\xi$ GPD model (continuous line). On the bottom panel, they are compared to the H0 exponential model. It appears that, in the first case ($\xi \neq 0$), sample estimates and model estimates are very close. Moreover, the sample estimates are always included in the corresponding 95 % confidence intervals. Let us also highlight that the confidence intervals obtained with the GPD approach are smaller than those obtained with the GEV approach since the GPD estimates are based on more points. The fit of the exponential distribution ($\xi = 0$) seems to be slightly worse for large return periods. The estimated 100-year return levels are reported in Table 1. They are compared to these obtained with the probability weighted moments estimation of the GPD parameters.

To summarize, excluding the results obtained with the exponential distribution (POT approach, $\xi = 0$), we end up with five estimations of the 100-year return level ranging from 1209 (block maxima, $\xi = 0$) to 1341 (POT, $\xi \neq 0$). Besides, the confidence intervals displayed on Figs. 1 and 2 provide an assessment of the uncertainty of each individual estimation. It appears that each of these five estimations belongs to the four 95 % confidence intervals computed with the other methods. This highlights the consistency between the estimations. However, we do not have any assessment of the global uncertainty, i.e. including the uncertainty linked to the choice of estimation method.

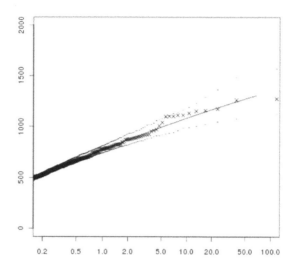

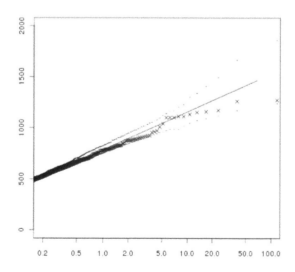

**FIGURE 2:** Return level plot (POT approach, maximum likelihood estimators). Top: assumption $\xi \neq 0$, bottom: assumption $\xi = 0$. Horizontally: $-1/\log(1 - 1/N)$, logarithmic scale, vertically: estimated N-year return level. Continuous line: estimation using the GPD, crosses: sample points, dots: 95 % confidence interval.

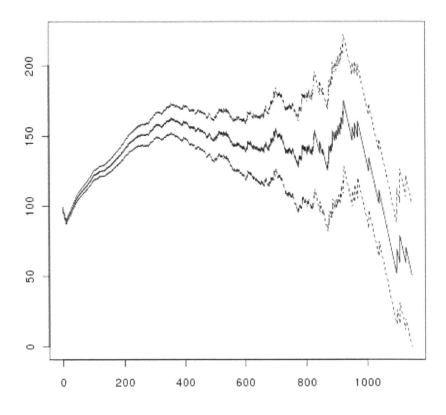

**FIGURE 3:** Mean residual life plot and associated 95 % confidence interval. Horizontally: threshold, vertically: mean excess function.

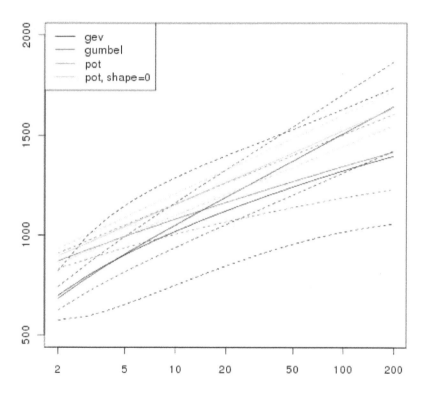

**FIGURE 4:** Overall findings with confidence intervals. Horizontal: return times. Vertical: estimated flow rate.

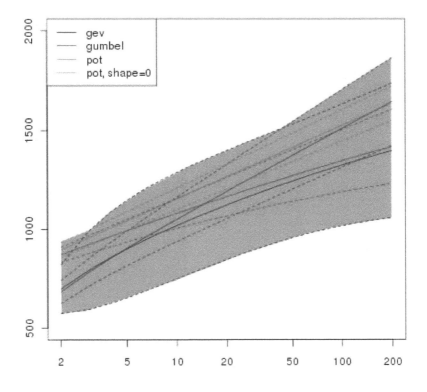

**FIGURE 5:** With 95 % confidence, the hazard is located in the blue zone. Horizontal: return times. Vertical: estimated flow rate.

## 7.6 GUIDING ACTION

Results show that it is difficult in the considered situation to obtain accurate reliable flow rates for rare or exceptional events. We can see that flow-rate estimates with a 95 % confidence interval (Figs. 4 and 5) vary between 920 m³ s⁻¹ (GEV lower limit) and 1767 m⁻³ s⁻¹ (Gumbel—upper limit).

Therefore, a risk management policy based merely on controlling the hazard is just not possible for Besançon. Risk management of rare events must be integrated in regions where work needs to be done on adapting the issues at stake and urban systems.

### 7.6.1 INITIATING STRATEGIC REFLECTION

Questions clearly need to be raised on the strategy for implementing re-silience. This strategy can be defined as the art of directing and coordinat-ing actions for attaining an objective. Strategic reflections cover analysis, decision-making and strategic action. Strategic reasoning seems to be the appropriate solution for implementing resilience as it enables the com-plex nature of urban elements and resources to be integrated into the data concerning the problem to be solved. Having or not having sufficient re-sources available can seriously influence the way objectives are defined. Reflections on resources also concern the virtues of what already exists and on the means of benefitting from them.

Development of an integrative approach to the strategic reflection concept is important for implementing resilience. It enables strategic re-flection to be envisaged as a global training and strategy development process that is comprised of interacting stages of analysis, decision-mak-ing and action.

"If analysis can be considered to be the quintessence of reflection, we must also consider that action is a form of reflection in itself.

Taking action means adapting, modelling and transforming intellectual concepts (decisions) into results that can be materially exploited depending on the conditions encountered when they are implemented. Under these con-ditions, action includes reflection; it is a form of reflection" (Torset, 2005).

Strategic reflection is based on analyses; it is nourished with, and formalized by, decisions and is enriched or renewed by action. It then offers a homogeneous frame of analysis for building up a strategy, from initial strategic notions through to the results obtained by actions.

- Strategic analysis: Strategic analysis is developed by characterizing regions requirements during and after crises on the one hand, and on the basis of the regions' resources and capacities on the other.
- New knowledge on modes of resilience and its organizational tools can be obtained by analysing already encountered situations. Innovation factors for strengthening resilience are also a source of information for the analysis.
- Strategic reflection (at a tactical level) concerns decision-making and tools for decision-making (and will also concern sustainability assessments for the strategies proposed at present.
- Action (at an operational level) is materialized by experimenting and debate on the evaluation of results.

As far as strategic analysis is concerned, the resilience of urban systems passes via specific approaches centred on smaller scales. The strategic analysis should help understand the hazards for the city and also the importance of critical infrastructure in the urban operation.

## 7.6.2 URBAN COMPONENT TYPOLOGY

The first action concerns the material components of an urban system as they play a crucial role before, during and after the crisis. Protection objectives must also be defined depending on the role played by the different urban components during flooding. Tools and methods of analysis now enable us to improve the way we can identify and locate these urban components and their functions (Prévil et al., 2003). In the Besançon catchment, three types of urban component have been identified where efforts must be made for designing a more resilient city (Fig. 6):

- urban components of a strategic nature, such as emergency centres, the gendarmerie and the town hall whose function is to shelter the persons who will be managing emergency situations and to provide logistical and institutional support during the crisis;
- urban components of an aggravating nature such as classified installations for environmental protection, hydrocarbon storage centres, etc. Should they

fail, these component elements will increase risks. It is important to know these component elements and take action beforehand to avoid the consequences of an initial disruption becoming any more serious due to a domino effect (for example, pollution resulting from non-protected stocks, industrial accidents, etc.);

- urban components of a minimizing nature: for example, refuges guaranteeing better resilience. These components generally offer protection against the risks and disruptions in which they are involved, but they can also generate risks or undergo important damage which will make emergency and post-crisis management less effective. In this way, spatialization, simulations and 3-D views can facilitate the way in which inherited or potential vulnerabilities are taken into account when defining urban projects. These tools also provide information on the flow rate at which urban components are liable to be flooded. Material measures can then make them less vulnerable.

## 7.6.3 APPROACHING RESILIENCE VIA URBAN SYSTEMS

For defining resilience objectives other than those concerning components, reflections must also be made on the way cities operate. Present-day technical urban networks are highly vulnerable; they possess great potential for suffering from damage. They are also sources of vulnerability on the scale of the urban system, as the way the city operates largely depends on the fact they operate satisfactorily. Two important and interconnected notions can be highlighted by analysing the behaviour of urban technical networks:

- the critical infrastructure notion where "critical" is synonymous with "essential" or "vital". A critical infrastructure can be defined as a set of installations and services that are necessary for the city (ASCE, 2009) to operate: their failure is a menace for the safety, economy, life style and public health of a city, a region or even a state;
- the notion of network interdependence: most critical infrastructures interact with each other. These interactions are often complex and unrecognized, because they go beyond the limits of the system in question.

Two types of interaction can be singled out when analysing interdependent critical infrastructures:

- interactions within a single critical infrastructure (energy, sewerage or road network);

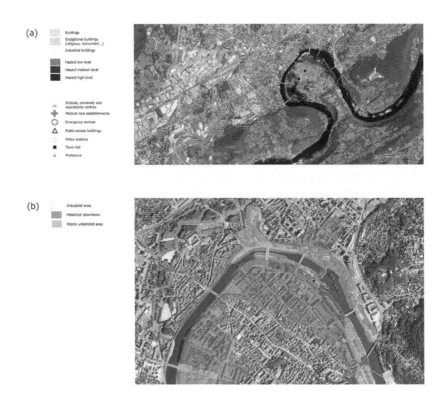

**FIGURE 6:** (a) Map of urban components; (b) city center land use in hazard area.

- interactions between different critical infrastructures (McNally et al., 2007), which requires a network of networks to be analysed (macro-network).

The least failure can have a knock-on effect on the whole system (Robert et al., 2009; Robert and Morabito, 2009; Serre, 2011). Therefore, analysis of interdependencies requires scales to be changed in order to analyse the component elements of a system (fine-scale) followed by the relations between different systems (a wider meta-system scale): a critical infrastructure is initially analysed as a system in itself and then, on a more widely encompassing scale, as a system of critical infrastructures (macro-network). A conceptual Spatial Decision Support System model is required for analysing the resilience of these technical systems (Balsells Mondejar et al., 2013). This model is based on three capacities (Fig. 7):

- The capacity for resisting a disruption resulting from material damage to networks following a hazard. The more a technical system is materially damaged, the more probable it will be that the system will dysfunction globally and the more difficult it will be to put it back into service. Operating reliability notions provide methods of determining damage to the system and taking account of interdependencies.
- The capacity to absorb a disruption, which depends on the alternatives that the network can offer following the failure of one or more of its component elements. For example, when a transport network is damaged, traffic will be transferred to routes that are alternatives to the initial itinerary. The more different routes there are, the less the disruption will be felt (Gleyze and Reghezza, 2007). These are alternatives that enable service continuity to be maintained and the network to operate in degraded mode. Methods resulting from the graph theory provide interesting answers.
- The capacity to recover, which is essential for a system to be resilient. For a network, recovery may simply be the time required for putting a damaged component back into service. In this case, purely technical aspects are conjugated with more organizational aspects. Recovery also concerns the accessibility of services needed for putting the network and any potentially damaged components back into service. The aim is to use spatial elements of analysis rather than organizational elements that require a great deal more information: recovery capacity assessments can be made with the help of geographic information sciences.

Strategic reflection could make decisions which will then be translated into action. This decision could concern a panel of return period and also estimating extreme flow rates using the extreme-value methods presented above and statistical uncertainty. This also implies setting those strategies in a long-term sustainable development context where societies will have to learn to live with natural disasters within their local area.

## 7.6.4 APPROACHING BY MEANS OF AWARENESS

Urban and industrial development in risk areas are kept under control by means of regulations. State policy is also based on prevention aspects. This principally concerns fostering a culture of risk: how can education and re-membrance make local inhabitants aware of a proven risk? Responsibility for reducing flood risks involves a common culture shared between State servic-es, the mayor and local authorities, public bodies, associations and citizens.

Work must be done in common not only for developing collective awareness of the causes, but also, and above all, for creating the collective and individual actions that need to be set up for protecting human life and reducing the vulnerability of services and property.

In Besançon, State services have created an online database for stor-ing references (press articles, photographs, plans, etc) on historic floods and making them available to the public. Anyone can consult the database and even add new elements to it. An Internet site acting as a flood observatory, which contains a certain number of documents and map-based initiatives, is also available. On the scale of the Doubs River watershed, exhibitions give local inhabitants information and a booklet containing texts from 1910, as well as postcards are also available. They show how inhabitants managed to organize themselves, relying on mutual solidarity both for lighting and heat-ing, or even for crossing water- filled streets and transporting fresh supplies to isolated persons. The city centre was isolated and the only way of trans-porting people outside the loop made by the Doubs River and back inside was via the bridge-keeper's shuttle system. The extent to which day-to-day life was hindered for two days is clearly visible, as well as the time needed to return to normal. What would the effects of this flood be today, taking into account our increasing vulnerability? In 1910, it was stocks of wood that settled under the La République bridge. Today, we would most certainly find other equally troublesome products jamming the river: wrecked cars, tanks and containers, as well as all sorts of other debris.

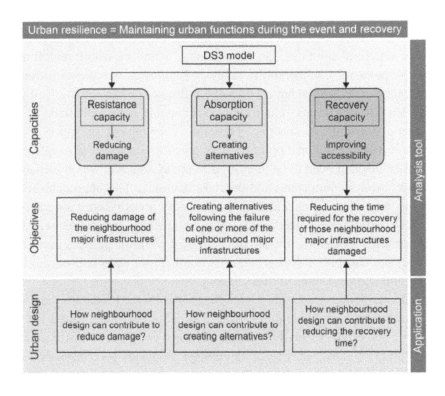

**FIGURE 7:** Application of conceptual Spatial Decision Support System model to the neighborhood level (Balsells Mondéjar et al., 2013).

## 7.7 CONCLUSIONS

An approach to uncertainty in hazard evaluation via different mathematical models for extreme values makes us aware of our knowledge status and guides our reflections as to how to implement resilience measures. When the level of uncertainty is important, which is the case for Besançon, it would appear to be an error and economically impossible to envisage resilience just by keeping the hazard under control by means of heavy structural solutions. On the contrary, using the whole region as a starting point for a vulnerability analysis enables us to recreate different geographical levers that have a decisive influence on risk situations: links between different scales, time frames, participants' roles and interests in a dynamic, non-static perspective. Therefore, carrying out an "autopsy" on resilience from a regional point of view presupposes the need to question the priorities that need to be identified in the system and which influence the way the system operates and the risks that exist. This approach should enable us to identify, characterize and classify areas where vulnerability is created and disseminated within a given territory. It is fundamental to concentrate on these areas when developing prevention policies inasmuch as they are capable of disrupting, compromising or even interrupting the operation and development of a territory.

The resilience strategy is a complement to hazard reduction and anticipation strategies. It requires risk to be actively appropriated by the persons involved, especially local populations, and for preventive actions (surveillance, alerts, etc.) to be developed alongside protection actions. Under these conditions, urban planning cannot be separated from an organizational dimension.

## REFERENCES

1.  Adger, W. N., Dessai, S., Goulden, M., Hulme, M., Lorenzoni, I., Nelson, D. R., Naess, L. O., Wolf, J., and Wreford, A.: Are there social limits to adaptation to climate change?, Clim. Change, 93, 335–354, 2008.
2.  ASCE: Guiding Principles for the Nation's Critical Infrastructure, ISBN 978-0-7844-1063-9, 42 p., 2009.

3.  Balsells Mondéjar, M., Barroca, B., Amdal, J. R., Diab, Y., Becue, V., and Serre, D.: Analysing urban resilience through alternative stormwater management options: application of the conceptual Spatial Decision Support System model at the neighbourhood scale, Water Sci. Technol., 68, 2448–2457, 2013.

4.  Barroca, B.: Risque et vulnérabilités territoriales – Les inondations en milieu urbain, University of Marne-la-Vallée, PhD thesis, 340p., 2006 (in French).

5.  Barroca, B., Bernardara, P., Mouchel, J. M., and Hubert, G.: Indicators for identification of urban flooding vulnerability, Nat. Hazards Earth Syst. Sci., 6, 553-561, doi:10.5194/nhess-6-553-2006, 2006.

6.  Bernardara, P., Schertzer, D., Sauquet, E., Tchiguirinskaia, I., and Lang, M.: Flood probability distribution tail: how heavy is it?, Stoch. Environ. Res. Risk Assess., 22, 107–122, 2008.

7.  Birkmann J., Cardona O. D., Carreno M. L., Barbat A. H., Pelling M., Schneiderbauer S., Kienberger, S., Keiler, M., Alexander, D., Zeil, P., and Welle, T.: Framing vulnerability, risk and societal responses: the MOVE framework, Nat. Hazards, 67, 193–211, 2013.

8.  Ceresetti, D., Ursu, E., Carreau, J., Anquetin, S., Creutin, J. D., Gardes, L., Girard, S., and Molinié, G.: Evaluation of classical spatial-analysis schemes of extreme rainfall, Nat. Hazards Earth Syst. Sci., 12, 3229–3240, doi:10.5194/nhess-12-3229-2012, 2012.

9.  Coles, S.: An Introduction to Statistical Modeling of Extreme Values, Springer Series in Statistics, Springer, 2001.

10. EEA: Impacts of Europe's changing climate – 2011 indicator-based assessment European Environment Agency, 2011.

11. Gardes, L. and Girard, S.: Conditional extremes from heavy-tailed distributions: An application to the estimation of extreme rainfall return levels, Extremes, 13, 177–204, 2010.

12. Gaume, E., Livet M., Desbordes M., and Villeneuve, J.-P.: Hydrological analysis of the river Aude, France, flash flood on 12 and 13 November 1999, J. Hydrol., 286, 135–154, 2004.

13. Gleyze, J.-F. and Reghezza, M.: La vulnérabilité structurelle comme outil de compréhension des mécanismes d'endommagement, Géocarrefour, 82, 17–26, 2007 (in French).

14. Klein, R. J. T., Nicholls, R. J., and Frank, T.: Resilience to natural hazards: how useful is this concept?, Environ. Hazards, 5, 35–45, 2004.

15. Lang, M., Perret C., Renouf E., Sauquet E., and Paquier A.: Incertitudes sur les débits de crue, Colloque SHF valeurs rares et extrêmes de précipitations et de débits pour une meilleure maîtrise des risques, Lyon, Publications SHF, 2006 (in French).

16. McNally, R. K., Lee, S.-W., Yavagal, S., and Xiang, W.-N.: Learning the critical infrastructure interdependencies through an ontology-based information system, Environ. Plann. B, 34, 1103–1124, 2007.

17. Milman, A. and Short, A.: Incorporating resilience into sustainability indicators : An example for the urban water sector, Glob. Environ. Change, 18, 758–767, (2008).

18. Neves, C. and Fraga Alves, M. I.: Reiss and Thomas' automatic selection of the number of extremes, Comput. Stat. Data Anal., 47, 689–704, 2004.

19. Payrastre, O.: Faisabilité et utilité du recueil de données historiques pour l'étude des crues extrêmes de petits cours d'eau. Etude du cas de quatre bassins versants affluents de l'Aude, Thèse de l'Ecole Nationale des Ponts et Chaussées, p. 202, 2005 (in French).
20. Payrastre, O., Gaume, E., and Andrieu, H.: Use of historical data to assess the occurrence of floods in small watersheds in the French Mediterranean area, Adv. Geosci., 2, 313–320, doi:10.5194/adgeo-2-313-2005, 2005.
21. Prévil, C., Thériault, M., and Rouffignat, J.: Analyse multicritère et SIG pour faciliter la concertation en aménagement du territoire: vers une amélioration du processus décisionnel?, Cahiers de géographie du Québec, 47, 35–61, 2003 (in French).
22. Robert, B. and Morabito, L.: Réduire la vulnérabilité des infrastructures essentielles, Lavoisier 2009 (in French).
23. Robert, B., Pinel, W., Pairet, J.-Y., Rey, B., and Coeugnard, C.: Organizational Resilience – Concepts and evaluation method, Montréal: Centre Risque & Performance, 2009.
24. Romero-Lankao P. and Dodman D.: Cities in transition: transforming urban centers from hotbeds of GHG emissions and vulnerability to seedbeds of sustainability and resilience: Introduction and Editorial overview, Curr. Opin. Environ. Sustain., 3, 113–120, 2011.
25. Romero Lankao, P. and Qin, H.: Conceptualizing urban vulnerability to global climate and environmental change, Curr. Opin. Environ. Sustain., 3, 142–149, 2011.
26. Serre, D.: La ville résiliente aux inondations Méthodes et outils d'évaluation, p. 173, Université Paris-Est, 2011 (in French).
27. Serre, D. and Barroca, B.: Preface "Natural hazard resilient cities", Nat. Hazards Earth Syst. Sci., 13, 2675–2678, 2013.
28. Torset, C.:La réflexion stratégique, objet et outil de recherche pour le management stratégique ?, 14. conférence internationale de management stratégique, Angers, France, juin 2005. (in French)
29. Zevenbergen, C., Cashman, A., Evelpidou, N., Pasche, E., Garvin, S. L., and Ashley, R.: Urban Flood Management, London, UK: Taylor and Francis Group, 2011.

# PART III

# ENERGY RESILIENCE

# CHAPTER 8

# Resilience, Sustainability and Risk Management: A Focus on Energy

BENJAMIN MCLELLAN, QI ZHANG, HOOMAN FARZANEH, N. AGYA UTAMA, AND KEIICHI N. ISHIHARA

## 8.1 INTRODUCTION

Energy is one of the keys to the development of nations and society—one of the vital lifelines that keeps societies and economies functioning. Civilization is dependent on a constant, consistent supply of energy; globally the demand for energy has been increasing consistently in parallel with growth in population and economic consumption [1]. However, the hazards associated with energy are also important to recognize—the risks associated with technology as well as the risks associated with the cutting of energy supply.

Given the widespread impact and publication of the accident at the nuclear power station in Fukushima, the risk of nuclear power has again been brought into the limelight. However, each of the alternative energy systems also has risks—some far less obvious than others. For example, some of the first images of earthquake damage in Japan on March 11th

_Resilience, Sustainability and Risk Management: A Focus on Energy._ © McLellan B, Zhang Q, Farzaneh H, Utama NA, and Ishihara KN. Challenges *3,2 (2012), doi:10.3390/challe3020153.* Licensed under Creative Commons Attribution 3.0 Unported License, http://creativecommons.org/licenses/by/3.0/.

were of fires and explosions occurring at LPG storage facilities at an oil refinery [2]. Moreover, it has been identified that, although renewable energy technologies such as solar panels may not have been damaged by the earthquake or tsunami, in the aftermath in many cases they continued to operate, even when the converters may have been damaged, thus creating an additional electrical hazard while not providing useful electricity [3].

This paper aims to build a framework for the consideration of risks associated with energy systems and to highlight failures in the aftermath of the March 11 disasters of 2011 that should ideally be avoided in future. The paper examines key energy systems in a variety of configurations and applications in order to present a comprehensive overview of the failures and advantages of the different alternatives. Furthermore, the consideration of risk is broadened to include alternative sustainability-related risks that are regularly ignored or under-examined in technological assessments of risk [4]. In particular, the paper addresses the potential beneficial contribution of energy systems in times of natural disaster.

Risk assessment and evaluation approaches in the design phase of processes and energy systems has been common practice since the late 1970's or early 1980's in many countries [5,6,7]—however, many current large scale power plants were designed prior to this. In terms of safety, there is empirical evidence that such assessments have reduced loss of life and infrastructure [8,9,10]. Some work has been done regarding infrastructure more generally with regards to climate change adaptation and the potential additional natural disaster risks that may be anticipated [11]. However, these approaches have largely focused on the operation's internal infrastructural integrity rather than seeking to incorporate sustainability risks, although some consideration to environmental risk has been given [12]. There is an opportunity, or indeed an urgent need, to use such structured techniques to identify risks and mitigation options beyond the plant gate— a boundary which is partially bridged by this study.

### 8.1.1 RESILIENCE AND VULNERABILITY

Resilience and vulnerability are terms utilized as approximate antonyms in this study. The interrelations and origins of research in the fields of resilience and vulnerability will not be covered in depth, as they have been widely reviewed and discussed elsewhere—e.g., [13,14]. However, characteristics of resilience have been summarized as [15]:

1. The amount of change the system can undergo and still retain the same controls on function and structure.
2. The degree to which the system is capable of self-reorganization to accommodate external changes.
3. The ability to build and increase the capacity for learning and adaptation.

On the other hand, vulnerability has been defined as having components that include [16]:

1. Sensitivity to perturbations or external stresses
2. The (lack of) capacity to cope or adapt
3. Exposure to perturbations.

The context of the current study is on energy systems and their impact on sustainability—especially when exposed to natural disasters. Therefore, the definition of vulnerability in this case is "susceptibility of the energy system, or component thereof, to damage due to a natural disaster, resulting in loss of key functions or significant negative impact on sustainability". Resilience is then defined as "the ability of an energy system, or component thereof, to withstand damage due to a natural disaster, to have a benign or negligible impact on sustainability in case of damage or loss, and to contribute in a positive way to societal recovery post-disaster."

In regards to the context of the present work, there have been a number of studies that have analyzed factors of resilience in energy systems. The geographical characteristics affecting the resilience—both in terms of vulnerability and response/regeneration time—are discussed in terms of the electrical distribution system, highlighting particular factors of importance such as population density and land use [17]. Electricity systems at the national scale have also been quantitatively analyzed using a set of resilience indicators that include such items as "carbon intensity" and "diversity" of generation, but this again has not delved specifically into the technological aspects of energy systems and does not explicitly address sustainability concerns [18]. Synergistic opportunities of integration of waste and energy [19] and of water networks [20] have highlighted the balance between efficiency and resilience (integration often improving efficiency, but potentially adding to vulnerability).

### 8.1.2 SUSTAINABILITY

Sustainability is a term that has been widely used and interpreted in a variety of ways. In most cases the definitions revert to some form of expression of the importance of preserving, enhancing and balancing the triple-bottom-line (TBL) of environment, economy and society. Often the "Brundtland" definition of sustainable development (SD) is utilized as a default: "development that meets the needs of the present without compromising the ability of future generations to meet their own needs" [21]. In the current work, we will largely follow this trend, taking sustainability to be a state of dynamic interplay between environment and society that ultimately contributes positively to indefinite human development and wellbeing whilst not overdrawing natural resources or over-burdening the environment in an irreversible manner. Rather than applying a TBL approach however, we use the "five capitals" framework (natural, social, human, manufactured and economic capital) [22] as the basis of our assessment. In this framework, the capitals refer to:

1.  Natural–the natural environment, including all environmental services and environmental quality
2.  Social–the networks or organizations that connect individuals
3.  Human–the individual's characteristics and wellbeing—notably skills, education and health
4.  Manufactured–the built environment and infrastructure
5.  Economic–monetary transactions and wealth.

Sustainability research has significant links to work in the field of resilience and vulnerability. Some examples of the connections have been in research on sustainable cities, which highlight the vulnerability of critical infrastructure as an important factor in the design of sustainable cities and the resilience of those cities [23]. Specifically, analysis has been conducted into the synergies between infrastructure planning for the reduction of risk in the event of terrorist or technological disasters and the potential reduction of environmental impacts [24]. The ecological, social and economic connectivity of the global society, and its vulnerability to environmental disruptions, have also been examined [25]. Likewise, the specific implications of environmental degradation in exacerbating natural disasters have been reviewed in light of the Indian Ocean Tsunami, showing some strong connections between the ultimate impact of the disaster and the vulnerabilities induced by policy decisions [26]. Some interesting work on full-cost accounting that applied the four capitals (excluding economic) has also tried to delve into the broader costs of coastal disasters [27]—although the focus was not specifically energy and therefore less detailed than the current approach.

Sustainability of energy is also an area that has been focused on by academics, politicians and industry across the world—e.g., [28,29,30,31]. The application of both centralized and decentralized renewable energy systems has been one specific focus for the integrated analysis of energy-sustainability-resilience. Renewable energy as a distributed energy technology providing diversity of generation and enhancing local skills and employment has been examined in detail [32,33], while the challenges of creating an entirely independent energy region have been shown to be particularly high [34], and may not be ideal in terms of resilience across annual fluctuations of renewable energy sources. On the larger scale, hydropower development in China as a renewable energy source has been highlighted as introducing a mixture of vulnerability and resilience due to the potential impacts on a large (often transboundary) community and environment and the necessity of good governance and stakeholder engagement [35].

Sustainability and resilience are often the topic of post-construction analysis and assessment. However, in terms of ability to improve operational performance in both safety and sustainability, it is important to consider the earlier stages of development [36]. Sustainability in design or "design for sustainability" (DfS) seeks to incorporate considerations of sustainability into the system even before it is constructed—which is often the most economically feasible timing as well as giving the highest sustainability impact and reducing risk [4,37,38]. Various frameworks and tools have been developed for measuring and incorporating sustainability [39], but these approaches are generally considered from the perspective of normal operations rather than from unusual conditions such as natural disasters (which are mostly considered in safety or risk management studies). Two such DfS approaches are considered here as an introduction to the potential implications on safety and resilience of energy systems in natural disasters of using a sustainability framework for design.

**TABLE 1:** Principles of "Green Engineering" [40] and considerations of resilient energy systems.

| | Principles | Resilience considerations for energy |
|---|---|---|
| 1 | Designers need to strive to ensure that all material and energy inputs and outputs are as inherently nonhazardous as possible. | Non-hazardous flows do not pose an exacerbated threat in a disaster situation. |
| 2 | It is better to prevent waste than to treat or clean up waste after it is formed. | Reducing waste produced in regular operations reduces storage of waste and potential for release in a disaster. |
| 3 | Products, processes and systems should be "output pulled" rather than "input pushed" through the use of energy and materials. | Energy systems driven by output will likely be readily ramped down or up subsequent to disaster-related demand shift. |
| 4 | Targeted durability, not immortality, should be a design goal. | Durability of infrastructure needs to correspond to potential disasters. |
| 5 | Design for unnecessary capacity or capability (e.g., "one size fits all") solutions should be considered a design flaw. | Flexible operation however, can be useful in disaster situations. |
| 6 | Design of products, processes and systems must include integration and interconnectivity with available energy and materials flows. | Utilization of locally available energy and materials may enhance resilience when supply lines are cut further afield. |
| 7 | Products, processes and systems should be designed for performance in a commercial "afterlife". | Subsequent to disaster, in the worst case infrastructure should be reusable for alternative applications. |
| 8 | Material and energy inputs should be renewable rather than depleting. | Renewable inputs are likely to be less hazardous and may rely less on long supply chains–although they may be vulnerable in some disaster situations. |

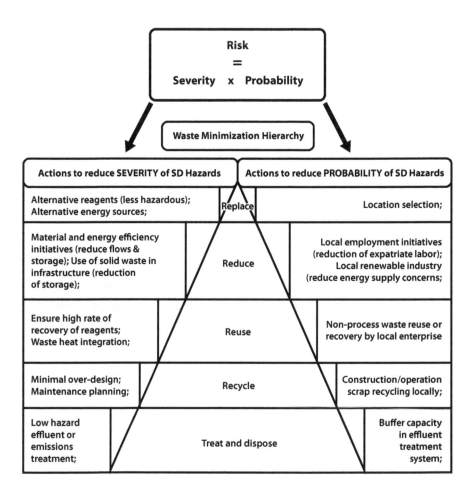

**FIGURE 1:** Waste minimization hierarchy and its relationship to reducing the severity and probability of hazards.

The first approach is "Green Engineering", with its 12 principles [40] to assist in improving the environmental performance of industrial operations or processes (of which the eight directly relevant principles are given in Table 1). It is apparent from the resilience considerations for energy that correspond to these principles, that in many cases there is a direct correlation between the design for environment and the resilience of the system in disasters.

The second framework considered is the Waste Minimization Hierarchy (WMH), which is shown in Figure 1. This figure gives some examples of initiatives that correspond to each of the five levels of the WMH and that have benefits in reducing either the severity or probability of a hazard, which in turn reduces the overall risk to environment and society from an operation. This approach to reducing environmental impacts from waste can also thus be utilized in improving energy system robustness in the face of natural disasters.

It is apparent from these two methodologies that there is strong linkage and potential between the use of sustainability principles, tools or frameworks in the design of energy systems and the development of more resilient systems. The framework applied in the remaining assessment of risk and resilience is the "five capitals" framework [22] which was described earlier.

## 8.2 METHODOLOGY

The current work employs a systematic methodology for identifying risks, vulnerabilities and opportunities for their mitigation. This methodology has been adapted from earlier work that applied a similar assessment to minerals industry projects [4]. The focus is broader than just technical risk, seeking to identify sustainability risks using the "five capitals" framework [22] to break down the sustainability components of energy systems. The examination specifically focuses on the vulnerability of different energy systems to natural disasters, and the subsequent impact or benefit that the energy system can have in disaster situations. Because the study examines energy systems under unusual circumstances of natural disasters, it does not take into account the standard operational impacts which are the consideration of most of the standard literature on risk and the externalities of energy [41,42,43].

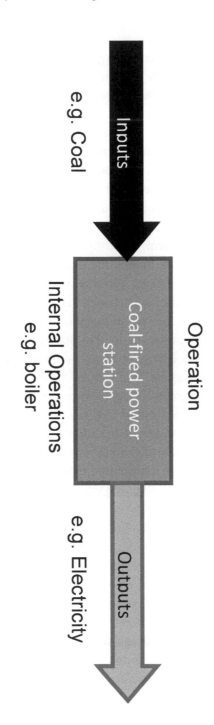

**FIGURE 2:** Input-operation-output diagram for the risk assessment.

The alternative energy systems considered in this assessment are representative of the key modern energy sources—both conventional fossil fuels and the more recently expanding renewable energy technologies. In this case, the energy systems are not analyzed down to the unit process level, as might be the case with a more rigorous technical assessment that might occur during the design or licensing of an operation—using techniques such as Hazard and Operability (HAZOP) studies for example [44,45]. Rather, in this study, we examine only the major flows and operational conditions that are identified as posing a particularly high risk by virtue of the magnitude of their volume, toxicity or energy content. Being focused on sustainability, the study also examines the flow-on effects of damage to the energy system.

The methodology first breaks the energy system down into inputs, outputs and internal operations—in this way covering the key aspects of the supply chain and emissions/wastes that could be affected by a natural disaster (see Figure 2). The assessment also includes the fuel extraction operations as a separate section but not the facilities that produce the equipment used in the energy systems—although these can certainly be affected by natural disasters. The specific flows and operations that were identified using this framework, and subsequently examined, are shown in Table 2 for the energy generating facilities and Table 3 for the fuel extraction and transportation stages of the supply chain. In the following step, the vulnerabilities of each of the key inputs, outputs and internal operations are analyzed through a structured HAZOP-like process. (In this case, we apply a subset of HAZOP guide words and a "grey box" model of the energy systems—focusing only on high influence operations to identify hazards.) Subsequently, the resulting potential impacts of natural disasters damaging the energy system are listed against the five capitals framework.

In most considerations of resilience, there is a focus on the system as a whole rather than individual elements as isolated nodes [46]. However in this case, we focus largely on the key energy generating equipment, as the vulnerability of such elements is the most crucial component and in most cases the most hazardous. However, in order to examine some of the systemic impacts and vulnerabilities, in addition to the identification of technology-specific risks, general configurations of energy systems and the ability for alternative systems to enhance societal resilience in the face of disaster are also examined. Furthermore, two key components–labor and the electricity grid–are considered separately due to their influence on every energy system. Finally, this analysis draws on significant examples from the March 11th disaster in Japan in 2011 to highlight the practical reality of energy systems in disaster situations.

**TABLE 2:** Inputs-internal operations-outputs for alternative energy systems and sub-components.

| Energy system | Inputs | Internal operations | Outputs |
|---|---|---|---|
| Nuclear power | Nuclear fuel | Reactor | Electricity |
| | Water | Cooling system | Steam and heated water |
| | Electricity | Steam cycle | Spent fuel |
| | | Spent fuel storage | |
| Coal-fired power | Coal | Boiler | Electricity |
| | Water | Steam cycle | Steam |
| | Oxidant (Air) | Tailings storage | Flue gas |
| | | Coal storage | Ash |
| Natural gas-power | Natural gas | Combustion turbine | Electricity |
| | Water | Steam cycle | Steam |
| | Oxidant (Air) | Fuel storage | Flue gas |
| Natural gas–heat/ fuel | Natural gas | Fuel storage/transmission | Heat |
| | Oxidant (Air) | Combustor/engine | Flue gas |
| Oil-power | Oil | Combustion engine | Electricity |
| | Water | Fuel storage | Flue gas |
| | Oxidant (Air) | | Heated water |
| Oil–heat/fuel | Oil | Fuel storage | Heat |
| | Oxidant (Air) | Combustor/engine | Flue gas |
| Hydro power | Water | Dam | Electricity |
| | | Turbines | Water |
| Geothermal | Water | Steam cycle | Water |
| Wind | Air | Turbine | Electricity |
| | | | Air |
| Solar-PV | Sunlight | PV panels | Electricity |
| | | Batteries (optional) | |
| Solar-Thermal | Sunlight | Thermal collector | Hot water |

**TABLE 3:** Extraction and transportation stages considered for the assessment.

| Energy system | Extraction | Transportation |
|---|---|---|
| Uranium | Mining | Road |
| | In-situ leaching | Ship |
| | Tailings storage | |
| Coal | Mine | Rail/conveyor |
| | Tailings storage | Ship |
| | Coal storage | |
| Natural gas | Well extraction | Pipeline |
| | Fuel storage | Ship |
| Oil | Well extraction | Pipeline |
| | Fuel storage | Ship |
| Biomass | Harvesting | Road/Rail |

## 8.3 RESULTS

The described methodology was applied to the flows and operations in Table 2 and Table 3. The resultant detailed responses to the HAZOP-type assessment are provided in the Appendix (Table A1, Table A2, Table A3, Table A4). From these responses, the general resilience and vulnerability statements associated with hazard categories that are associated with most energy systems are summarized in Table 4, while Table 5 then highlights the key risks from each of the energy systems on the five capitals, based on the worst case scenario for a variety of natural disasters. The particular disasters to which given systems were considered to be most susceptible given the technology, location and recent examples, are also indicated in the far right hand column. Some general categories of vulnerability, common to most energy systems—labor, supply chains and the electricity grid, are discussed further in section 4, and are therefore not elaborated on in these tables. The first row of Table 5 further highlights some of the key impacts on the five capitals that arise from given natural disasters.

**TABLE 4:** System components vulnerability and resilience.

| Component | Vulnerability | Resilience |
|---|---|---|
| Fuel supply chain | Long supply chain–import facility or transport loss exposure | Long supply chain less vulnerable to simultaneous damage |
| | Short supply chain–vulnerable to simultaneous damage | Short supply chain can allow continued or rapid-recovery of operations |
| Fuel storage | Onsite storage can be damaged and cause fire, explosion, contamination or other health and environmental hazards | Onsite storage can enable continued operation despite supply-chain loss. Frequency of fuel delivery directly related to onsite storage capacity and usage rate: |
| | | Nuclear fuel–infrequent delivery makes damage unlikely; |
| | | Hydrocarbon fuel–sufficient onsite storage allows operation during recovery; |
| Fuel characteristics | Flammable or explosive fuels provide high energy but increase hazard of storage; | Solid fuels are less likely to disperse widely; |
| | Liquid fuels likely to pool or disperse in waterways; | Gaseous fuels will disperse widely and rapidly; |
| Utilities supply to plant | Water for cooling system and steam cycles: | Electricity vital for many plant operations–mostly self-supplied so that operations can continue if grid supply is lost; |
| | Sudden loss of water can cause damage; | |
| | Nuclear plant without water can fail in some cases, causing damage to plant; | |
| | Electricity loss at nuclear power plant particularly hazardous; | |
| | Sudden electricity loss may damage other thermal plant, but more likely to cause sudden loss of reactive conditions; | |

**TABLE 4:** *Cont.*

| Component | Vulnerability | Resilience |
|---|---|---|
| Waste storage | Onsite storage can be damaged and dispersed: <br><br> Large volume, low level waste such as ash from coal or bio-mass or <br><br> Low volume, high level waste such as spent nuclear fuel | |
| Non-fuel energy sources | Wind–vulnerable to overly windy (e.g., storm or cyclone conditions); <br><br> Water–hydro and ocean energy can be damaged by spikes in influx; <br><br> (Hydro dam damage most critical hazard) <br><br> Water–hydro power is dependent upon water inflow; <br><br> Sunlight–solar power vulnerable to overcast conditions associated with storm activity; panels vulnerable to hail and high winds; | Non-fuel sources except water are only temporarily interruptible–short term recovery is likely; <br><br> Geothermal heat is unlikely to spike or trough; <br><br> Dam construction is often designed specifically for flood mitigation and can reduce the scale and intensity of negative impacts of high rainfall. |
| Electricity output | Grids vulnerable to flooding and high wind particularly; <br><br> Lack of output has serious societal impact; <br><br> Larger grids lead to higher impact; <br><br> Continued output to damaged grid is hazardous to health and equipment; | Continued output can support societal recovery; <br><br> Most generation facilities can rapidly decrease generation–but restarting after long lay-off is most difficult for nuclear or coal; |

**TABLE 5:** Major risks of impact on the five capitals of generic natural disasters and alternative energy systems.

| | Human | Social | Economic | Manufactured | Natural |
|---|---|---|---|---|---|
| Disaster direct impact on capitals | Health and life | Support networks | Loss of revenue | Loss of facilities | Loss of populations or habitats |
| | Access to facilities | Consumer loss | Cost of repair | Damage to facilities | Loss of regular environmental services |
| | Skills loss | | Flow-on loss of productivity | | |
| System impacts on capitals subsequent to natural disaster | | | | | Disaster susceptibility |
| Biomass | Health– minor fire risk | Moderate impacts associated with loss of electricity | Moderate impacts associated with loss of electricity | Minor fire risk | Bush fire; Flooding; Gale force winds; |
| Coal | Health– fire or contamination | Major impacts associated with loss of electricity | Major impacts associated with loss of electricity | Fire risk | Bush fire; Earthquake; Flooding; Gale force winds; Tsunami; |
| Gas | Health– fire | Moderate impacts associated with loss of electricity | Moderate impacts associated with loss of electricity | Fire or explosion risk | Bush fire; Earthquake; Flooding; Gale force winds; Tsunami; |
| Geothermal | Health– minor contamination possible | Minor loss of electricity | Minor loss of electricity | - | Minor contamination possible | Earthquake; flooding |

**TABLE 5:** *Cont.*

|  | Human | Social | Economic | Manufactured | Natural | |
|---|---|---|---|---|---|---|
| Hydro | Health–dam break flash flooding risk; | Minimal to major impacts associated with loss of electricity; | Minimal to major impacts associated with loss of electricity and flooding; | Loss of infrastructure by flash flooding; | Silt and relocated water pattern impact on ecology; | Earthquake; Severe flooding; |
|  |  | Dislocation–flooding destruction; | Loss of workforce; Loss of revenue; Loss of consumer base; |  |  |  |
| Nuclear | Health–long term contamination | Dislocation–long term contamination | Loss of workforce | | Long term contamination | Earthquake; Tsunami; |
|  | Loss of employment |  | Loss of consumer base | | | |
| Oil | Health–fire or contamination | Moderate impacts associated with loss of electricity | Moderate impacts associated with loss of electricity | Fire or explosion risk | Moderate medium term impacts on local species from fire or contamination; | Bush fire; Flooding Earthquake Tsunami |
| Solar | Health–electrocution risk | Minor impacts associated with loss of electricity | Minor impacts associated with loss of electricity; Cost of repair could be significant proportional to supply; | - | - | Flooding; Tsunami; Gale force winds; |
| Tidal/wave |  | Potential shipping hazard; | Minor to moderate impacts associated with loss of electricity; | Equipment coming loose may become an additional floating hazard; | - | Tsunami; |

**TABLE 5:** *Cont.*

|  | Human | Social | Economic | Manufactured | Natural |
|---|---|---|---|---|---|
|  |  | Minor to moderate impacts associated with loss of electricity; |  |  |  |  |
| Wind | - | Minor impacts associated with loss of electricity | Minor impacts associated with loss of electricity; | Turbine towers may topple, causing damage to nearby capital; | Minor fire risk; | Gale force winds; |

## 8.4 DISCUSSION

It can be seen from the results shown in the previous section (Table 4 and Table 5) that, in general, non-fuel-based renewable energy technologies are less hazardous—both during operation and in a natural disaster situation—largely due to the lower intensity of operating conditions. The exception is hydropower, which under most conditions would support resilience but may, under particular circumstances such as earthquakes or exceptionally high rainfall, add to a disaster through dam failure and consequent flooding. However, renewable energy technologies have vulnerabilities that may make them particularly susceptible to poor performance during natural disasters. For example, wind turbines must be shut off during high winds, making them useless in storms or cyclones—although they were shown to perform well in the March 11 earthquake and tsunami [47]. Likewise, solar photovoltaics may not produce sufficient electricity in overcast conditions that would also be associated with storm or rainfall-associated disasters.

The following subsections highlight some of the major specific impacts of energy systems in natural disasters.

## 8.4.1 GENERAL FACTORS AFFECTING VULNERABILITY

Some characteristics of energy systems may be generalized across all alternative systems—in particular, the configuration of these systems as centralized or decentralized. Although there are many discussions as to what exactly defines a decentralized or distributed system (some useful definitions elsewhere [48,49]), in general, such systems should provide energy to a relatively small number of users in a limited geographic area. Typical of decentralized energy systems would be small scale generators supplying rural communities or solar panels providing energy for a single home. At the other end of the scale, centralized energy systems rely on economy of scale, with large scale production that is distributed over a wide area for numerous users—examples include typical modern nuclear and coal-fired power plants connected to the electricity grid. Table 6 highlights some of the key impact factors and the important aspects of time and diversity that can support or adversely affect resilience.

Damage to infrastructure can present many risks that are unassociated with the infrastructure's use in the energy system—for example, hazards associated with falling objects or drowning. Such risks are not considered in detail here, but may be associated with loss of operating conditions, loss of labor, loss of control of the system and thereby exacerbate the natural disaster.

Two key factors in the resilience of energy systems can be highlighted as: diversification and prioritization. Diversification is typical of distributed energy systems that utilize renewable energy produced by a variety of small to medium-sized generators [32]. The key resilience factor of diversity is that it is less likely for multiple sources of energy to be cut off or damaged than for a single source. However, the level of skills, care and maintenance applied to larger, centralized systems may also be an advantage for preventing damage in a disaster. The larger, centralized systems also typically gain priority from government and corporate stakeholders who wish to see the largest consumer base supply recovered the quickest [17]. This can mean that the smaller generators take longer to be repaired and may also present a hazard to health and environment for a longer time.

**TABLE 6:** Factors affecting risk and vulnerability of centralized or decentralized energy systems.

| Configuration | Impact factors | Time | Diversity |
|---|---|---|---|
| Centralized | Larger number of dependent users; | Larger infrastructure leads to longer delay for reconstruction; | Low diversification–vulnerable to specific feedstock loss; |
| | Typically larger scale storage of feedstocks and waste; | Priority often given for reconstruction because of larger user base; | |
| | Typically more intense operating conditions (temperatures and pressures); | | |
| | Wide-spread impact; | | |
| Decentralized | Localized impact; | Shorter delay to start-up; | Higher diversification–more robust to loss of single feedstock; |
| | Smaller number of users; | Lower-priority for reconstruction in many cases; | |

## 8.4.2 LABOR RISKS

Energy systems cannot be considered separate from the society they support. One key non-infrastructure component of energy systems that may be generalized across many alternative systems is the labor input. Small-scale renewable energy systems are generally autonomous, with labor only required for installation and maintenance—however, for all other systems some level of human control is inevitable. Natural disasters can be alleviated or exacerbated by the response of those in control—for recent examples: flood mitigation in south-east Queensland (January 2011) and the Fukushima nuclear power plant. However, an additional consideration is the ability of the labor force to attend work to keep the energy system operating. In the case of a natural disaster, access to installations is often hampered, and it may fall to those operating the plant at the time of the disaster to continue working for extended periods until access and labor relief is achieved. This risk was particularly exposed by the nuclear accident at Fukushima, in which new laws needed to be passed so that operators could continue to work at the plant beyond their previous yearly limit of radiation exposure. The ability to maintain a sufficient workforce with the requisite skills will be boosted by operations requiring low onsite staffing, but with large local or regional workforce. This is likely to benefit sustainability from a socio-economic perspective during normal operations as well as facilitating recovery from natural disasters. Decentralized renewable energy technologies may be favored from this perspective [32,33]—especially in cases where ongoing deployment of technologies has enabled the growth of skills for maintenance within the area [37].

### 8.4.3 TRANSMISSION AND DISTRIBUTION NETWORK RISKS

Under the current paradigm of energy generation and distribution, large scale, centralized production feeds extensive networks of users via the electricity grid. Grid damage presents a risk to all centralized energy systems—and subsequently to their customers. (Moreover, grid electricity loss is an additional hazard in nuclear systems that require an external energy supply for safety during shutdown.) Whilst most modern emergency facilities, hospitals and some other commercial or government facilities may have back-up electricity supplies (on-site generator), the onsite storage of fuel is limited and may not be sufficient if transportation infrastructure is also damaged. Above ground infrastructure for suspending transmission lines is susceptible to high winds and strong flood waters in particular, while underground electrical infrastructure can be damaged by flooding [11]. Damaged energy distribution networks can also pose a hazard to health—electrification or fire risk.

### 8.4.4 SUPPLY CHAIN IMPACTS

Significant both in terms of general operational fuel cycle impacts and in terms of vulnerability to natural disasters are the supply chain elements from extraction of fuel and distribution of fuel to electricity generating facilities. The analysis of supply chain vulnerabilities in Table A4 highlighted a number of key points. Firstly, the mining or extraction of fuels and its subsequent impact on the energy system is largely related to the distance of separation between the extraction and usage, and the frequency of delivery. Natural gas delivered by pipeline directly into a distribution network or coal delivered to a mine-gate power plant is vulnerable to immediate and prolonged disruption of electricity generation. By contrast, the mining of uranium or disruption to coal transported to distant power plants is likely to have a lower or delayed impact, due to the infrequency of delivery.

In general, thermal coal is shipped relatively short distances, as is biomass. This implies that energy systems using these fuels are more likely to be simultaneously impacted by a natural disaster, whereas nuclear, oil or liquefied natural gas plants are likely to use fuels of more distant origin and therefore less likely to be simultaneously interrupted. The impacts of natural disasters are in most cases likely to be similar for the extraction, delivery and generation stages of the life cycle, although potentially the extraction phase will involve larger volumes of material—e.g., a whole mine or an oil well compared to a single oil or coal storage facility.

## 8.4.5 CONTRIBUTION IN TIMES OF NATURAL DISASTER

In natural disasters, swift and safe search, rescue and recovery efforts are highly dependent on available energy resources. This is true both for organized emergency response and for independent household recovery. Without fuel, emergency generators, vehicles and equipment cannot run. Without electricity, communications, coordination and evacuation facilities are impotent. Therefore it is imperative that energy systems are resilient and contribute safely to the ongoing provision of energy in the event of a natural disaster. However, onsite storage or continued generation in the case of damaged facilities may contribute to exacerbating the disaster by creating an electrical or fire hazard. Some key criteria that would delineate resilient and non-resilient energy systems are discussed briefly in this section.

In order to support communities hit by natural disasters, energy systems must be:

1.  Continuous—operating safely throughout or restarting safely immediately post-disaster
2.  Robust—not easily damaged in case of potential natural disasters
3.  Independent—able to operate for a continuous period (in the order of days to weeks) post-disaster without relying on physical intervention from outside (local source of energy or sufficient storage, and with an appropriate local skills base to operate)
4.  Controllable—able to be readily shut-down or with output adjusted depending on conditions
5.  Non-hazardous—able to provide energy in a way that does not cause an additional unwarranted hazard
6.  Matched to demand—able to provide energy in the form and quantity that is needed, in the location it is needed, when it is needed

Continuous operation is closely linked to robustness and controllability, in that the physical impact of a given natural disaster on the energy system impacts the ability of the system to continue to operate and its controllability under non-ideal operating conditions. This robustness is related both to the mode of the disaster and the physical location, materials, support structures, mode of operation and disaster defenses of the energy system. For example, solar photovoltaics are often roof-mounted, making them less vulnerable to flood damage but more vulnerable to high winds. Controllability of the system under normal and abnormal operating conditions is often the result of initial design, and can be particularly affected by the choice of electronic or physical control. Often in response to disaster, energy systems can only be controlled in a single direction or across a small band of operating conditions—e.g., wind power increases until the cut-out speed, when the turbine is stopped; nuclear power automatically shuts down.

The importance of independent generation is never more starkly highlighted than in times of natural disaster. Many energy systems can run relatively autonomously under normal conditions—with only external monitoring and occasional maintenance. However, if generators are shut down for safety prior to a disaster, or if they sustain damage due to the disaster, then external intervention is typically necessary—especially when remote operation becomes impossible due to grid or communication network damage. Independence for a period of days is often possible with relatively low storage capacity onsite. However, longer periods require access to energy sources locally or through the reopening of fuel supply routes. Independence is also closely tied to matching with demand. In most modern societies, electricity, oil and gas are the most ubiquitous energy sources—electricity for power applications and oil and gas for heating, cooking and for transportation. Natural disasters can often cause links in the supply chain to be cut. Due to the inflexibility of most household energy structures, and the "lock-in" to electricity for appliances or petroleum for transportation, the designed end use becomes inoperable without an external fuel supply of the required type (e.g., conventional gasoline cars will not run on electricity, and neither will electric stoves run on gas). Matching demand in situations of natural disaster may require the ability to operate flexibly in the production of alternative energy carriers (which, for instance, may be a benefit of fossil fuel-based energy systems or of hydrogen energy systems [50]) or having sufficient spare capacity to expand generation to cover for loss elsewhere in the system.

Finally, the aspect of being non-hazardous (either inherently, or by control) is of key importance in disasters. Of the examined energy systems, perhaps the least hazardous is geothermal, given that it relies on relatively low heat, and that heat is a more readily detectible hazard than electricity for humans. Furthermore, geothermal energy systems can be readily shut down to avoid leakage. Wind power and hydro will in most circumstances also be relatively non-hazardous. On the other hand, the storage of fuels in nuclear and thermal power plants presents an inherent risk, which is only mitigated by sound infrastructure design and control.

## 8.5 CONCLUSIONS

One of the key requirements of energy systems that promote resilient societies in the face of natural disasters is that the energy system itself is resilient to natural disasters. This paper has examined various existing energy systems from the perspective of hazards posed to and by the energy system in the case of natural disasters.

The current work highlights the connection between sustainability and resilience in the design of energy systems. It is specifically highlighted that frameworks for improving the operational sustainability performance of energy systems can also provide benefits of reducing risks in times of natural disaster. The five capitals model of sustainability is also applied to help identify the risks beyond the plant boundary.

As a result of the assessment, six key criteria for energy systems to contribute to the resilience of a community in the face of natural disasters are highlighted, indicating that energy systems should be: (1) Continuous; (2) Robust; (3) Independent; (4) Controllable; (5) Non-hazardous; and (6) Matched to demand. Energy systems that correspond to these criteria will contribute to sustainable development in both standard operation and in times of extraordinary hardship.

Regarding these criteria, geothermal energy is regarded as one of the technologies that most contributes to societal sustainability. Non-renewable technologies tend to involve hazardous materials, while renewable energy systems may be hazardous if uncontrolled. Supply of energy from local sources is positive for rapid recovery but may mean that all stages of the supply chain are damaged simultaneously, in which case distant sources of energy may be preferred from a vulnerability perspective.

## REFERENCES

1.  IEA, World Energy Outlook 2008; International Energy Agency: Paris, France, 2008.
2.  COSMO Oil Co. Ltd. Lpg tanks fire extinguished at chiba refinery (5th update). Available online: http://www.cosmo-oil.co.jp/eng/information/110321/index.html (accessed on 26 March 2011).
3.  Kurokawa, K. Exploiting all the possibilities of PV power generation. In From Post-Disaster Reconstruction to the Creation of Resilient Societies; Keio University: Tokyo, Japan, 2011.
4.  McLellan, B.C.; Corder, G.D. Risk reduction through early assessment and integration of sustainability in design in the minerals industry. J. Clean. Prod. 2012.
5.  Kletz, T.A. Hazop-past and future. Reliab. Eng. Syst. Saf. 1997, 55, 263–266.
6.  Kletz, T.A. The origins and history of loss prevention. Process Saf. Environ. Prot. 1999, 77, 109–116.
7.  Swann, C.D.; Preston, M.L. Twenty-five years of hazops. J. Loss Prev. Process Ind. 1995, 8, 349–353.

8.  Kletz, T. Incidents that could have been prevented by hazop. J. Loss Prev. Process Ind. 1991, 4, 128–129.
9.  Groves, W.A.; Kecojevic, V.J.; Komljenovic, D. Analysis of fatalities and injuries involving mining equipment. J. Saf. Res. 2007, 38, 461–470.
10. Pitblado, R. Global process industry initiatives to reduce major accident hazards. J. Loss Prev. Process Ind. 2011, 24, 57–62.
11. Engineering the Future, Infrastructure, Engineering and Climate Change Adaptation-Ensuring Services in An Uncertain Future; The Royal Academy of Engineering: London, UK, 2011; p. 107.
12. García-Serna, J.; Pérez-Barrigón, L.; Cocero, M.J. New trends for design towards sustainability in chemical engineering: Green engineering. Chem. Eng. J. 2007, 133, 7–30.
13. Vogel, C.; Moser, S.C.; Kasperson, R.E.; Dabelko, G.D. Linking vulnerability, adaptation, and resilience science to practice: Pathways, players, and partnersh. Glob. Environ. Chang. 2007, 17, 349–364.
14. Turner, B.L., II. Vulnerability and resilience: Coalescing or paralleling approaches for sustainability science? Glob. Environ. Chang. 2010, 20, 570–576.
15. Wardekker, J.A.; de Jong, A.; Knoop, J.M.; van der Sluijs, J.P. Operationalising a resilience approach to adapting an urban delta to uncertain climate changes. Technol. Forecast. Soc. Chang. 2010, 77, 987–998.
16. Manuel-Navarrete, D.; Gómez, J.J.; Gallopín, G. Syndromes of sustainability of development for assessing the vulnerability of coupled human-environmental systems. The case of hydrometeorological disasters in central America and the Caribbean. Glob. Environ. Chang. 2007, 17, 207–217.
17. Maliszewski, P.J.; Perrings, C. Factors in the resilience of electrical power distribution infrastructures. Appl. Geogr. 2012, 32, 668–679.
18. Molyneaux, L.; Wagner, L.; Froome, C.; Foster, J. Resilience and electricity systems: A comparative analysis. Energy Policy 2012, 47, 188–201.
19. Kharrazi, A.; Masaru, Y. Quantifying the sustainability of integrated urban waste and energy networks: Seeking an optimal balance between network efficiency and resilience. Procedia Environ. Sci. 2012, 13, 1663–1667.
20. Li, Y.; Yang, Z.F. Quantifying the sustainability of water use systems: Calculating the balance between network efficiency and resilience. Ecol. Model. 2011, 222, 1771–1780.
21. World Commission on Environment and Development, Our Common Future; Oxford University Press: Oxford, UK, 1987; p. 347.
22. Forum for the Future The five capitals. Available online: http://www.forumforthefuture.org/project/five-capitals/overview (accessed on 12 September 2011).
23. Branscomb, L.M. Sustainable cities: Safety and security. Technol. Soc. 2006, 28, 225–234.
24. Coaffee, J. Risk, resilience, and environmentally sustainable cities. Energy Policy 2008, 36, 4633–4638.
25. Kissinger, M.; Rees, W.E.; Timmer, V. Interregional sustainability: Governance and policy in an ecologically interdependent world. Environ. Sci. Policy 2011, 14, 965–976.

26. Srinivas, H.; Nakagawa, Y. Environmental implications for disaster preparedness: Lessons learnt from the indian ocean tsunami. J. Environ. Manag. 2008, 89, 4–13.
27. Gaddis, E.B.; Miles, B.; Morse, S.; Lewis, D. Full-cost accounting of coastal disasters in the united states: Implications for planning and preparedness. Ecol. Econ. 2007, 63, 307–318.
28. Afgan, N.H.; Gobaisi, D.A.; Carvalho, M.G.; Cumo, M. Sustainable energy development. Renew. Sustain. Energy Rev. 1998, 2, 235–286.
29. Kinrade, P. Toward a sustainable energy future in australia. Futures 2007, 39, 230–252.
30. Lee, S.-C.; Shih, L.-H. Enhancing renewable and sustainable energy development based on an options-based policy evaluation framework: Case study of wind energy technology in taiwan. Renew. Sustain. Energy Rev. 2011, 15, 2185–2198.
31. Lior, N. Sustainable energy development (May 2011) with some game-changers. Energy 2012, 40, 3–18.
32. O'Brien, G.; Hope, A. Localism and energy: Negotiating approaches to embedding resilience in energy systems. Energy Policy 2010, 38, 7550–7558.
33. del Río, P.; Burguillo, M. An empirical analysis of the impact of renewable energy deployment on local sustainability. Renew. Sustain. Energy Rev. 2009, 13, 1314–1325.
34. Schmidt, J.; Schönhart, M.; Biberacher, M.; Guggenberger, T.; Hausl, S.; Kalt, G.; Leduc, S.; Schardinger, I.; Schmid, E. Regional energy autarky: Potentials, costs and consequences for an austrian region. Energy Policy 2012, 47, 211–221.
35. McNally, A.; Magee, D.; Wolf, A.T. Hydropower and sustainability: Resilience and vulnerability in china's powersheds. J. Environ. Manag. 2009, 90, S286–S293.
36. Corder, G.D.; McLellan, B.C.; Green, S.R. Delivering solutions for resource conservation and recycling into project management systems through susop®. Miner. Eng. 2012, 29, 47–57.
37. Corder, G.D.; McLellan, B.C.; Bangerter, P.J.; van Beers, D.; Green, S.R. Engineering-in sustainability through the application of susop®. Chem. Eng. Res. Des. 2012, 90, 98–109.
38. McLellan, B.C.; Corder, G.D. Designing-in Sustainability in Industrial Projects and Processes. In Proceedings of EcoDesign 2011-7th International Symposium on Environmentally Conscious Design and Inverse Manufacturing, Kyoto, Japan, 30 November-2 December 2011; Matsumoto, M., Umeda, Y., Masui, K., Fukushige, S., Eds.; Springer: Kyoto, Japan, 2011.
39. McLellan, B.C.; Corder, G.D.; Giurco, D.; Green, S. Incorporating sustainable development in the design of mineral processing operations-review and analysis of current approaches. J. Clean. Prod. 2009, 17, 1414–1425.
40. Anastas, P.T.; Zimmerman, J.B. Design through the 12 principles of green engineering. Environ. Sci. Technol 2003, 37, 94–101.
41. Burgherr, P.; Hirschberg, S. Severe accident risks in fossil energy chains: A comparative analysis. Energy 2008, 33, 538–553.
42. Inhaber, H. Risk with energy from conventional and nonconventional sources. Science 1979, 203, 718–723.
43. Ricci, P.F.; Cirillo, M.C. Health risks analysis of energy systems: Issues and approaches. Environ. Int. 1984, 10, 367–376.

44. Crawley, F.; Preston, M.; Tyler, B. Hazop: Guide to Best Practice: Guidelines to Best Practice for the Process and Chemical Industries; Institution of Chemical Engineers: London, UK, 2000.
45. Kletz, T.A. Hazop and Hazan: Identifying and Assessing Process Industry Hazards; The Institution of Chemical Engineers: London, UK, 1999.
46. Fiksel, J. Sustainability and resilience: Towards a systems approach. Sustain. Sci. Pract. Policy 2006, 2, 14–21.
47. Ushiyama, I. Activities on Offshore Wind Power Generation in Japan. In Proceedings of World Ocean Forum, Busan, Korea, 26-28 October 2011.
48. Ackermann, T.; Andersson, G.; Söder, L. Distributed generation: A definition. Electr. Power Syst. Res. 2001, 57, 195–204.
49. Alanne, K.; Saari, A. Distributed energy generation and sustainable development. Renew. Sustain. Energy Rev. 2006, 10, 539–558.
50. Afgan, N.; Veziroglu, A. Sustainable resilience of hydrogen energy system. Int. J. Hydrog. Energy 2012, 37, 5461–5467.

*There are several supplemental files that are not available in this version of the article. To view this additional information, please use the citation on the first page of this chapter.*

# PART IV

# PUBLIC HEALTH RESILIENCE

# CHAPTER 9

# Resilience to the Health Risks of Extreme Weather Events in a Changing Climate in the United States

KRISTIE L. EBI

## 9.1 INTRODUCTION

Public health has an impressive history of identifying and reducing health threats, as evidenced by the dramatic increase in life expectancy in the past 100 years. With more than 150 years of experience in identifying and responding to threats to health, public health should be well placed to address the health risks of climate change, including those posed by extreme weather and climate events (e.g., heatwaves, droughts, floods, windstorms). Initial policies and measures to adapt to the health impacts of climate change emphasized enhancing current health protection to climate sensitive health outcomes, such as implementing heatwave early warning systems [1]. Although such actions are vital, they will not necessarily reduce risks in a future climate; changing baselines mean that current efforts could be neutral or, in a worst case, maladaptive under new weather pat-

*Resilience to the Health Risks of Extreme Weather Events in a Changing Climate in the United States © Ebi KL.* International Journal of Environmental Research and Public Health *8,12 (2011),* doi:10.3390/ijerph8124582. *Licensed under a Creative Commons Attribuion 3.0 Unported License,* http://creativecommons.org/licenses/by/3.0/.

terns. Further modifications are needed to increase the resilience of public health and health care policies and measures in a changing climate, where resilience is defined as the ability to timely and effectively anticipate, prepare for, respond to, and recover from climate change and other risks; definitions of resilience vary across disciplines, with all incorporating elements of effectively responding to, limiting impacts from, and recovering after a perturbation. The key difference from current public health practice is in explicitly acknowledging climate change in the design, implementation, monitoring, and evaluation of strategies, policies, and measures. Actions to reduce vulnerability to extreme weather events may not promote adaptation and resilience if they do not specifically consider the risks of climate change and how those risks could change over time.

Responsibility for the prevention of climate-sensitive health risks rests with individuals, community and state governments, national agencies, and others, with the roles and responsibilities varying by health outcome [2,3]. Actions to increase resilience across all actors include top-down (such as strengthening and maintaining disaster risk management programs) and bottom-up (such as through programs to increase social capital) measures, and need to focus not only on the risks presented by climate change but also on the underlying vulnerabilities that determine the extent and magnitude of impacts today and in the future.

Public health in the U.S. has historically focused on top-down measures aimed at reducing individual exposures and behaviors that increase disease risk. This approach is a consequence of public health tending to view the world through a medical model, with individuals and communities more as units for intervention than as partners in action. This framing of health results in programs that focus on individual-level factors that increase or decrease risk of adverse health outcomes, where risks typically are either addressed by changing individual risk factors (e.g., tobacco smoking, high cholesterol, and high blood pressure) or by top-down interventions (e.g., regulations to control emissions of air toxics). However, a growing body of research has clearly shown that community and social factors are significant predictors for a range of health outcomes [4]. The result has been increasing awareness that communities need to be active partners in identifying and addressing a wide range of health risks if public heath programs are to be effective and sustainable [5].

**TABLE 1:** Relative public health impacts of selected extreme weather events. Source: Modified from [7].

| Public Health Impact | Storms | Floods | Heat | Drought | Wildfire |
|---|---|---|---|---|---|
| Number of deaths | Few | Few, but can be many in flash floods | Moderate to high | Few | Few to moderate |
| Number of severe injuries | Few | Few | Moderate to many cases of heat stroke | Unlikely | Few to moderate |
| Worsening of existing chronic illnesses | Widespread | Focal to widespread | Widespread | Widespread | Focal to widespread |
| Increased pests and vectors | Widespread | Widespread | Unlikely | Possible | Unlikely |
| Risk of an epidemic | Unlikely | Unlikely | Unlikely | Unlikely | Unlikely |
| Food scarcity | Uncommon | Uncommon | Unlikely | Common | Possible |
| Loss of clean water | Widespread | Focal to widespread | Unlikely | Widespread | Focal |
| Loss of sanitation | Widespread | Focal to widespread | Unlikely | Likely among displaced populations | Likely among displaced populations |
| Loss and/or damage of health care systems | Widespread | Focal to widespread | Unlikely | Unlikely | Focal |
| Loss of shelter | Widespread | Focal to widespread | Focal to widespread | Focal to widespread | Focal |
| Permanent migration | Unlikely | Unlikely | Unlikely | Likely | Unlikely |

A challenge to increasing public health resilience to the health risks of climate change is that no single organization or institution has comprehensive authority and responsibility for creating, maintaining, improving, and evaluating the Nation's public health infrastructure. Public health in the U.S. is promoted and delivered by a wide array of organizations and institutions, with highly variable resources, staffing, and performance capacity [6]. This is a consequence of a core value of the U.S. federal system: state and local autonomy and/or control, particularly in matters affecting personal risks and private interests. Each state is unique in its laws, regulations, governance, and budgets for public health, resulting in highly uneven abilities to deliver public health services.

There are significant opportunities for public health to increase the resilience of communities to the health risks of climate change. Mandates relevant to managing the risks of extreme weather and climate events are used as examples of opportunities for mainstreaming adaptation and increasing resilience, including (1) the all-hazards approach to health risks; (2) "healthy cities" programs; and (3) regionalization of public health services. Possible indicators for measuring public health preparedness and community resilience to extreme events also are discussed.

Keim [7] identified the wide range of possible public health impacts of extreme events expected to worsen with climate change (Table 1). The impacts experienced in a particular location will vary, depending on the effectiveness of the public health and health care systems, and on the vulnerabilities specific to that location.

## 9.2 ALL-HAZARDS APPROACH TO HEALTH RISKS

An all-hazards approach to health risks is increasingly a foundation for organizing public health efforts to prepare for emergencies. These efforts could be modified to take into consideration the health risks associated with extreme weather events, including that the magnitude, extent, and duration of many types of events are increasing with climate change [8]. To increase resilience, the all-hazards approach would need to specifically incorporate the risks associated with climate change, and would need to have a strong monitoring and evaluation orientation to ensure that current risk management policies and measures are regularly revisited to ensure that they continue to be effective as the frequency and patterns of extreme events change.

The all-hazards approach is based on two laws that recognized weaknesses in state and local public health systems. The Public Health Improvement Act of 2000 calls for a plan to assure the preparedness of every community in the nation. The 2006 update and reauthorization of the Public Health Security and Bioterrorism Act, now called the Pandemic and

All-Hazards Preparedness Act (Public Law No. 109-417) was intended "to improve the Nation's public health and medical preparedness and response capabilities for emergencies, whether deliberate, accidental, or natural." Funding is made available annually to meet a variety of goals, including developing and sustaining essential state, local, and tribal public health security capabilities, maintaining vital public health and medical services to allow for optimal operations in the event of a public health emergency, and addressing the needs of at-risk individuals.

A further mandate is the Department of Homeland Security National Preparedness Guidelines that define what it means for the nation to be prepared for any hazard [9]. The guidelines include national planning scenarios designed to focus contingency planning at all levels of government and in partnership with the private sector; included is a scenario for a major hurricane and for food contamination. The scenarios form the basis for coordinated Federal planning, training, exercises, and grant investments needed to prepare for emergencies. There is a universal task list with more than 1,600 unique tasks to facilitate efforts to avoid, prepare for, respond to, and recover from major events. In addition, 37 specific capabilities are listed that communities, the private sector, and all levels of government should collectively possess to respond effectively to disasters. The National Preparedness System then describes the actions needed at all levels of government, the private sector, nongovernmental organizations, and individuals to work together to achieve the priorities and capabilities outlined in the Guidelines. The Department of Homeland Security followed issuance of these guidelines with Presidential Directive 21 that established a National Strategy for Public Health and Medical Preparedness [10]. This directive focused on threat awareness, prevention and protection, surveillance and detection, and response and recovery, which are all highly relevant to emergency management of extreme weather events.

In response to these laws and other mandates, the U.S. Centers for Disease Control and Prevention (CDC) has developed performance measures, targets, definitions, instructions, data collection, and submission methods to demonstrate the degree of state, local, and tribal all-hazards preparedness [11]. Much of the data necessary to measure and monitor preparedness can be obtained during commonly occurring events, such as infectious disease outbreaks. In addition, state, local, and tribal entities who receive funding for all-hazards preparedness are expected to conduct drills and exercises to ensure that information is available for each performance measure. For example, one activity developed to support planning and local jurisdictions is the Cities Readiness Initiative to prepare major U.S. cities and metropolitan areas to effectively respond to a large-scale bioterrorist attack by dispensing antibiotics to their entire identified population within 48 hours of a decision to do so [12]. According to CDC, developing and maintaining the capacity to achieve this goal has enhanced communication and collaboration across state and local boundaries, and has helped local and state planners identify capabilities, strengths, and shortcomings through preparedness planning and technical assistance reviews. Because not all communities are involved, the overall level of preparedness can vary significantly within and across regions.

Based on performance measures and experiences, states have undertaken activities to improve their preparedness. For example, the Mississippi State Department of Health did not have the medical surge capacity to care for the thousands of individuals with special medical needs who were displaced during Hurricane Katrina [11]. All-hazards funding has since been used to equip buildings on selected campuses to act as special medical needs shelters for use in the event of a storm, a pandemic outbreak, or other natural or man-made disaster. In addition, alterations to electrical power systems will enable climate control and life support systems in the event of a power loss, and facilities are being retrofitted for use by physically challenged individuals. As evidence of progress, the Department of Health recently used the Mississippi Health Alert Network to notify the state's healthcare system of a serious outbreak of pertussis [11]. One person notified every participating physician, hospital, and other medical care provider in about six hours, with a verified delivery rate of 90%. Previously, such contact took a minimum of 12–14 hours with a 50% success rate. This capacity will be highly valuable in the event of another storm surge or other climate-related event.

CDC created a Local Public Health Preparedness and Response Capacity Inventory for rapid assessment of a public health agency's ability to respond to public health treats and emergencies [13]. The focus areas are preparedness, planning and readiness, surveillance, laboratory capacity, communication, and training. To assist local health departments in meeting preparedness mandates, the National Association of County and City Health Officials (NACCHO) developed a competency-based training and recognition program, Project Public Health Ready (PPHR) with three project goals; one is all-hazards preparedness planning [14]. The PPHR criteria are continuously updated to include the most recent federal initiatives. In addition, NACCHO provides a toolkit of best practices, self-assessments for local health departments, guidance on regional approaches to preparedness, and evaluations.

In addition to these activities, there are a growing number of methods and tools developed by a wide range of organizations to facilitate preparedness, including the Association of State and Territorial Health Officials (ASTHO), Council of State and Territorial Epidemiologists, National Association of Local Boards of Health, Association of Public Health Laboratories, Agency for Healthcare Research and Quality, and many others. One example is an online map-based planning tool developed by the Western New York Public Health Alliance Practice Center with its partners, the National Opinion Research Center at the University of Chicago and the Pennsylvania State University Center for Environmental Information (www.cei.psu.edu/evac). The tool assumes that when a disaster strikes an urban area, a significant number of residents will self-evacuate to surrounding rural communities. The tool allows users to select a city of interest and model how communities within a 150-mile radius might be affected by the spontaneous evacuation of urban residents following a dirty bomb explosion, chemical incident, or influenza pandemic. Maps show numbers of evacuees received by each surrounding county and their resulting population changes. Information on the number of hospital beds, hotel rooms, and other resources can be used to determine the resources a region may need to respond to evacuee and resident needs. Such tools could be modified to increase their relevance for other communities and in other contexts.

These and other initiatives are intended to improve public health preparedness to the types of impacts and disasters that are projected to increase with climate change, although they have yet to explicitly take climate change into account. For example, in 2001, some state public health departments did not have enough epidemiologists to investigate suspected anthrax cases, and had not fully anticipated the extent of coordination needed among first responders [11]. By 2007, in every state, cooperative agreements supported additional staff, public health departments had established relationships and conducted exercises with emergency management and other key players, and emergency operations centers were in place at CDC and almost all state public health departments to coordinate response activities. As noted earlier, these public health programs can be modified to specifically include the health risks of climate change, with strong monitoring and evaluation to ensure effectiveness as situations change.

However, an independent analysis of the Nation's preparedness for public health emergencies found that although significant progress has been made, there are important areas where continued and concerted actions are needed [15]. On a range of measures, federal and state polices still fall short of stated goals. Variation in preparedness among the states results in different levels of protection from potential emergencies. The report concluded there is a need to increase accountability, strengthen leadership; enhance surge capacity and the public health workforce; modernize technology and equipment; and improve community engagement. These are all determinants of adaptive capacity. For example, an evaluation of 24/7 critical case reporting that compared self-reports with operational data found that 18 of 19 local health department in a convenience sample indicated that they had a process in place to receive and respond to calls on a 24/7 basis, but operational assessments found that only two consistently met the prescribed standard [16].

A key problem is that assessments tend to focus on public health structures, but there is little evidence that links specific preparedness structures with the ability to execute effective responses to protect population health [17]. Focusing only on public health structures misses the additional factors that affect resilience, such as the social capital needed for effective community response to disasters and threats; programs need a broader basis that incorporates all relevant agencies and departments. It also misses changes that are needed to urban social-ecological systems, through urban planning and design standards for the built environment, to lower risks to individuals and communities when affected by an extreme event.

Modifying the goals and approaches of all-hazards programs to incorporate the changing risks associated with alterations in the frequency, intensity, and duration of extreme events, would increase preparedness today and under future climates.

## 9.3 REGIONALIZATION OF HEALTH SERVICES

Since 2001, there has been a national emphasis on increasing emergency preparedness for possible terrorist attacks by regionalizing local services, where regionalization is the addition of a regional structure to supplement local government agencies. Although the US relies on state and local health departments to provide essential public health services, few counties, cities, or towns are capable of responding to health emergencies on their own. For example, a national survey conducted in 2000–2001 found that only 25% of local public health jurisdictions felt they were able to deliver at least 60% of the essential public health services in the event of a terrorist attack [18]. In response, there has been increasing regionalization of public health services [19].

Rationales for regionalization include reducing socioeconomic and fiscal disparities between metropolitan and outlying areas [20]; creating economies of scale; and sharing human and financial resources regionally to improve efficiency and consistency of delivery of public health services, as well as reduce duplication of efforts, across local, regional, and state levels [21]. The potential benefits include improving both public health preparedness and response, as well as creating social capital. NACCHO, through its Project Public Health Ready program, identified four approaches to facilitating effective regionalization [14]: networking through the interactive sharing of preparedness information and plans between individuals and organizations; coordinating local health departments for planning events, such as training or exercises; standardizing across local health departments through mutual adoption of emergency preparedness functions; and centralizing resources for planning and response under a single entity, such as a single web portal, an emergency notification systems, or regional epidemiologic support.

Lessons learned from case studies of public health responses to outbreaks of West Nile virus, SARS, monkeypox, and hepatitis A in the U.S.

illustrate the value of regionalization to increase preparedness [22]. One lesson was that the required public health response is not necessarily proportional to the number of people actually exposed, infected or ill, or the number of deaths. This is because public health efforts to identify additional cases (particularly of cases of infectious diseases) through active surveillance are likely to result in the identification of many potential cases, including individuals who do not have the disease but are worried that they do. Extensive population-based prevention efforts, such as education campaigns, may be necessary to limit disease transmission and reduce the health consequences. These demands stress the capacity of public health systems even when the actual number of cases is small, such as a disease outbreak following a flooding event. Therefore, increased sharing of human and financial resources can increase the capacity of local public health agencies to cope when challenged with an outbreak or emergency.

Worldwide and national attention on increasing preparedness for pandemic influenza has increased interest in building regional surge capacity [21] including centralizing staff, space, and supplies. The lack of surge capacity was one of the current constraints to adaptation noted in the climate change adaptation assessment conducted by the Maryland Commission on Climate Change [23].

An example is Massachusetts, which regionalized its 351 autonomous cities and towns, each with its own local health department or board of health (as required by state law), into public health emergency preparedness regions and sub-regions to more effectively distribute CDC funding for state preparedness [21]. One action taken was to implement regional training programs, which advanced an iterative cycle of developing plans, training personnel, testing preparedness, and refining plans to clarify specific roles and responsibilities [24].

Regionalization is an opportunity to facilitate sharing of knowledge and resources to prepare for and respond to climate sensitive health threats. A challenge is that public health organizations and institutions tend to spend much of their time dealing with health threats as they occur. Chronic underfunding means that recognized gaps in preparedness often are not addressed until there are adverse health outcomes, after which there will be funding for a few years before funds are diverted to the next critical need [25].

Other challenges to increasing regionalization include that little is known about the best ways to define a region and what are the best practices [21]. Barriers to regionalization include fear of the loss of local autonomy and identity; lack of formal legal agreements that address liability and mutual aid; and resources to participate in regional activities. Public health agencies typically do not have command and control authority over important resources, such as hospitals, health care providers, and other government agencies, that are needed for an optimal public health response [22]. At the same time, there needs to be effective communication and coordination during a public health emergency. Jurisdictional arrangements can be complicated across state, regional, county, and city entities, each of which operates under different political leadership structures and local governmental and community organizations. Clearly, regionalizing public health services needs to address authority issues and complex jurisdictional arrangements. But addressing these challenges can increase resilience of communities and states to a range of risks, including extreme weather events associated with climate change.

## 9.4 HEALTHY CITIES PROGRAMS

Urban health assesses the impact of urban living on health through the physical environment, the social environment, and access to heath and social services [26]. Programs to promote urban health have generally focused on specific problem areas or populations. One widespread program is Healthy Cities, which is a community problem-solving process for health promotion. WHO defines a Healthy City [27] as "one that is continually creating and improving those physical and social environments and expanding those community resources which enable people to mutually support each other in performing all the functions of life and developing their maximum potential."

The Healthy City (now called Healthy Communities) process involves encouraging community participation, assessing community needs, establishing priorities and strategic plans, soliciting political support, taking local action, and evaluating progress [28]. The initial goals were to address primary health care and health promotion. It challenged communities to develop projects to reduce inequalities in health status and access to services, and to develop healthy public policies at the local level through a multi-sectoral approach and increased community participation in health decision-making. The basis of the European WHO Healthy Cities is that inequalities in social conditions (and therefore health) are unjustified and that their reduction should be an overriding public health objective [27]. Decreasing population vulnerability and building social capital provide a strong basis for increasing resilience to the health risks of extreme weather events.

The Healthy Cities programs typically view health broadly, and develop projects to collaboratively address the root causes of problems [27]. Projects are initiated independently, based on a range of philosophical orientations and with different sponsoring organizations and funding. CDC, NACCHO, and others provide information and tools for Healthy Cities projects. Common areas of action include the environment, community safety, immunizations, tobacco, and youth. Some of the characteristics that make a city a Healthy Cities include effective leadership; a multi-sectoral committee or steering group that directs the project; a strong economy; community participation; information used in citywide planning; needed technical support obtained; program viewed as a credible resource for health in the community; effective networking; and smaller city size or neighborhoods in large cities [28]. These factors also facilitate effective adaptation. Although some of these characteristics are beyond the control of the local community, many can be advanced through community partnerships. Given the variety of ways in which these characteristics can be measured, and the large number of interactions that could increase or decrease resilience, there is no consensus on the best indicators for designating a city as a "Health City".

Glouberman et al. [29] propose a framework that combines the urban health and Healthy Cities approaches into a Health in Cities process that recognizes the particular vulnerabilities and problems faced by specific populations within an urban environment and the effects of the urban environment on all city residents (i.e., that the health of a city is a multi-directional interaction between the residents and the environment). They advocate approaching health promotion in cities from the perspective of complex adaptive systems; that is, the entire system can be viewed as a network of relationships and interactions. This is a key component of adaptive management. Successful strategies for change in cities will not succeed without understanding the local context. Similar conclusions have been reached in assessing the adaptive capacity of communities to address the risks of climate change.

It is possible to build on the Healthy Cities or Health in Cities approaches to address the health risks of climate change [30]. Local actions aimed to reduce vulnerability and increase adaptive capacity can be designed to also reduce greenhouse gas emissions through alterations in transport policies and other actions [3]. Ebi and Semenza [5] outline a framework for facilitating community-based adaptation to the health impacts of climate change, with each step designed to enhance social capital and build resilience.

## 9.5 POSSIBLE INDICATORS OF COMMUNITY RESILIENCE TO EXTREME EVENTS

Assessing the ability of communities to prepare for and effectively respond to the health impacts of extreme events requires identification of a limited set of metrics for regular measurement and monitoring, either as mandated by the relevant agencies or by extending their mandate. General indicators of community resilience to extreme weather events associated with climate change can be categorized into (1) indicators of public health preparedness; (2) indicators of health-outcome specific morbidity and mortality; and (3) indicators of socioeconomic, political, technical, infrastructure, and other vulnerabilities.

English et al. [31] identified climate change and health indicators for the United States that were chosen to describe elements of environmental sources, hazards, exposures, health effects, and intervention and prevention activities. Some indicators are measures of environmental variables that can directly or indirectly affect human health, such as maximum and minimum temperature extremes, while others can be used to project future health impacts based on changes in exposure, assuming exposure-response relationships remain constant. Indicators were categorized into four areas: environmental, morbidity and mortality, vulnerability, and policy responses related to adaptation and greenhouse gas mitigation. These indicators cover measures that, if monitored, would track the magnitude and extent of health impacts.

Developing indicators for evidence-based measures of preparedness for a range of disasters is limited by the rarity of large-scale public health emergencies, and because other agencies typically have primary responsibility for emergencies. Using the ten essential public health services, CDC established a National Public Health Performance Standards Program to improve the quality of public health practice and the performance of public health systems [32]. The program is a collaborative effort of seven national partners that describes the public health activities that should be undertaken at state and local levels. It provides web-based tools and resources for state and local public health systems to self-assess and improve their performance. The local public health system assessment focuses on all entities that contribute to public health services within a community, while the local public health governance assessment focuses on the local governing bodies accountable for public health (i.e., boards of health, councils, or county commissioners). For state public health system assessments, there should be planning and implementation; state-local relationships; performance management and quality improvement; and public health capacity and resources for each essential service. For local public health governance, there should be oversight of the activities undertaken by local public health systems. The standards developed do not explicitly address the health risks of climate change, but easily could with the development of indicators for monitoring the risks posed by climate change and the effectiveness of programs and activities to increase current and future community resilience.

Indicators of key determinants of health outcomes need to include factors that influence the ways that social, environmental, behavioral, and health services determine population health [33]. Populations living in certain regions may have increased risks for specific climate-sensitive health outcomes due to their regions' baseline climate, abundance of natural resources such as fertile soil and fresh water supplies, elevation, dependence on private wells for drinking water, and/or vulnerability to coastal surges or riverine flooding [34]. Socioeconomic factors interact with the physical threats of extreme weather events and other disasters in determining vulnerability and sensitivity; they may increase the likelihood of exposure to harmful agents, interact with biological factors that mediate risk, and/or lead to differences in the ability to adapt or respond to exposures or early phases of illness and injury [34].

In addition to indicators of socioeconomic, physical, and other vulnerabilities to the health risks of climate change, there is growing recognition that communities are complex systems that are inadequately described in the traditional medical paradigm where risk factors have unidirectional associations with adverse health outcomes [35]. The dynamics of communities have been shown to determine health in their own right, independent of an individual's health and socioeconomic status. In addition, interventions that did not incorporate a broader understanding of the population in which they are being implemented (e.g., cultural context, social norms) have had limited success [36]. As in ecology, multidirectional relationships may more effectively describe the vulnerability and resilience of the population. Effectively addressing vulnerabilities to the health risks of climate change requires not only interventions to reduce specific health vulnerabilities, but also the societal, cultural, environmental, political, and economic context that increases vulnerability [37,38]. Engaging communities in the process of adaptation will not only enhance their resilience to climate stressors, but also may increase their ability to cope with a wide range of other societal issues [5], as evidenced by successes with the Healthy Cities programs.

## 9.6 CONCLUSIONS

Individuals and communities resilient to climate change anticipate risks; reduce vulnerability to those risks; prepare for and respond quickly and effectively to threats; and recover faster, with increased capacity to prepare for and respond to next threat. Public health institutions are structured to anticipate and prepare for risks, reduce identified vulnerabilities, and rapidly respond through a wide range of programs and activities, and have a long history of doing so effectively. However, preventing additional morbidity and mortality due to climate change will require modification of current and implementation of new programs and activities to increase resilience to climate change, taking into consideration the local context, including socioeconomic, geographic, the built environment, and other factors [2]. Programs designed for other purposes provide opportunities for increasing the capacity of public health organizations and institutions to avoid, prepare for, and effectively respond to the health risks of extreme weather and climate events.

Incorporating elements of adaptive management, a structured and iterative process of decision-making in the face of imperfect information, into public health practice will increase its effectiveness to address the health risks of climate change, through a strong and explicit focus on iteratively managing risks [39].

A critical challenge is to enhance the capacity of public health to increase its own and facilitate community resilience in less-than-robust environments [6]. Public health infrastructure has been seriously and systematically underfunded owing to its low priority among state and federal policy makers. A survey in the late 1990s, before passage of the Public Health Improvement Act and the Pandemic and All-Hazards Preparedness Act, found total per-capita spending on essential public health services of $37–$102 among local agencies and $86–$232 among state agencies [25]. There were subsequent improvements in public health funding, but funding is subject to changing priorities and is unevenly distributed across regions and issues of concern, leaving too many communities vulnerable to the extreme events associated with climate change.

Despite this, considerable improvement in community resilience can be expected as key public health agencies and organizations, such as the Department of Health and Human Services, CDC, NACCHO, ASTHO, the Red Cross, and others, facilitate the inclusion of future climate change risks into current and planned strategies and programs, such as those highlighted. In addition to traditional approaches to improving the public health infrastructure and systems, the social-ecological determinants of health should be explicitly included when identifying opportunities for increasing resilience to the health risks of climate change.

## REFERENCES

1. St Louis, M.E.; Hess, J.J. Climate change: Impacts on and implications for global health. Am. J. Prev. Med. 2008, 35, 527–538.
2. Ebi, K.L. Public health responses to the risks of climate variability and change in the United States. J. Occup. Environ. Med. 2009, 51, 4–12.
3. Frumkin, H.; Hess, J.; Luber, G.; Malilay, J.; McGeehin, M. Climate change: The public health response. Am. J. Public Health 2008, 98, 435–445.
4. World Health Organization Commission on the Social Determinants of Health. Closing the Gap in a Generation: Health Equity Through Action on the Social Determinants of Health; World Health Organization: Geneva, Switzerland, 2008; pp. 1–246.
5. Ebi, K.L.; Semenza, J. Community-based adaptation to the health impacts of climate change. Am. J. Prev. Med. 2008, 35, 501–507.
6. Baker, E.L.; Potter, M.A.; Jones, D.L.; Mercer, S.L.; Cioffi, J.P.; Green, L.W.; Halverson, P.K.; Lichtveld, M.Y.; Fleming, D.W. The public health infrastructure and our Nation's health. Ann. Rev. Public Health 2005, 26, 303–318.
7. Keim, M.E. Building human resilience: The role of public health preparedness and response as an adaptation to climate change. Am. J. Prev. Med. 2008, 35, 508–516.
8. Intergovernmental Panel on Climate Change (IPCC)Climate Change 2007: Synthesis Report; Contribution of Working Groups I, II and III to the Fourth Assessment Report of the Intergovernmental Panel on Climate Change; Core Writing Team-Pachauri, R.K., Reisinger, A., Eds.; IPCC: Geneva, Switzerland, 2007; pp. 1–104.
9. Department of Homeland Security (DHS). National Preparedness Guidelines; DHS: Washington, DC, USA, 2007. Available online: http://www.dhs.gov/xlibrary./National_Preparedness_Guidelines.pdf (accessed on 27 July 2011).
10. Department of Homeland Security (DHS). Homeland Security Presidential Directive 21: Public Health and Medical Preparedness; DHS: Washington, DC, USA, 2007. Available online: http://www.dhs.gov/xabout/laws/gc_1219263961449.shtm (accessed on 27 July 2011).
11. Centers for Disease Control and Prevention (CDC). Public Health Preparedness: Mobilizing State by State; A CDC Report on the Public Health Emergency Preparedness Cooperative Agreement; CDC: Atlanta, GA, USA, 2008. Available online: http://www.bt.cdc.gov/publications/feb08phprep/ (accessed on 27 July 2011).
12. Centers for Disease Control and Prevention (CDC). Cities Readiness Initiative; CDC: Atlanta, GA, USA, 2009. Available online: http://www.bt.cdc.gov/cri/ (accessed on 27 July 2011).
13. Costich, J.F.; Scutchfield, F.D. Public health preparedness and response capacity inventory validity study. J. Public Health Manag. Pract. 2004, 10, 225–233.
14. National Association of County and City Health Officials (NACCHO), Planning Beyond Borders: Using Project Public Health Ready as a Regional Planning Guidance for Local Public Health; NACCHO: Washington, DC, USA, 2007.
15. Levi, J.; Vinter, S.; St Laurent, R.; Segal, L.M. Ready or Not? Protecting the Public's Health from Diseases, Disasters, and Bioterrorism; Trust for America's Health, Robert Wood Johnson Foundation: Washington, DC, USA, 2008.

16. Dausey, D.; Lurie, N.; Diamond, A. Public health response to urgent case reports. Health Aff. (Millwood) 2005.
17. Nelson, C.; Lurie, N.; Wasserman, J. Assessing public health emergency preparedness: Concepts, tools, challenges. Ann. Rev. Public Health 2007, 28, 1–18.
18. Baker, E.L., Jr.; Koplan, J.P. Strengthening the Nation's public health infrastructure: Historic challenge, unprecedented opportunity. Health Aff. (Millwood) 2002, 21, 15–27.
19. Beitsch, L.M.; Kodolikar, S.; Stephens, T.; Shodell, D.; Clawson, A.; Menachemi, N.; Brooks, R.G. A state-based analysis of public health preparedness programs in the United States. Public Health Rep. 2006, 121, 737–745.
20. Mitchell-Weaver, C.; Miller, D.; Deal, R., Jr. Multilevel governance and metropolitan regionalism in the USA. Urban Stud. 2000, 5-6, 851–876.
21. Koh, H.K.; Elqura, L.J.; Judge, C.M.; Stoto, M.A. Regionalization of local public health systems in the era of preparedness. Ann. Rev. Public Health 2008, 29, 205–218.
22. Stoto, M.A.; Dausey, D.J.; Davis, L.M.; Leuschner, K.; Lurie, N.; Myers, S.; Olmsted, S.; Ricci, K.; Ridgely, M.S.; Sloss, E.M.; et al. Learning from Experience. The Public Health Response to West Nile Virus, SARS, Monkeypox, and Hepatitis A Outbreaks in the United States; Rand Corporation: Santa Monica, CA, USA, 2005; pp. 1–170.
23. Maryland Commission on Climate Change. Maryland's Climate Change Commission's Climate Action Plan; Maryland Commission on Climate Change: Baltimore, MD, USA, 2008. Available online: http://www.mdclimatechange.us/ (accessed on 27 July 2011).
24. Streichert, L.C.; O'Carroll, P.W.; Gordon, P.R.; Stevermer, A.C.; Turner, A.M.; Nicola, R.M. Using problem-based learning as a strategy for cross-discipline emergency preparedness training. J. Public Health Manag. Pract. 2005, November (Suppl), S95–S99.
25. Atchison, C.; Barry, M.A.; Kanarek, N.; Gebbie, K. The quest for an accurate accounting of public health expenditures. J. Public Health Manag. Pract. 2000, 6, 93–102.
26. Galea, S.; Vlahov, D. Urban health: Evidence, challenges, and directions. Ann. Rev. Public Health 2005, 26, 341–365.
27. Awofeso, N. The Healthy Cities approach—Reflections on a framework for improving global health. Bull. WHO 2003, 81, 222–223.
28. Flynn, B.C. Healthy Cities: Toward worldwide health promotion. Ann. Rev. Public Health 1996, 17, 299–309.
29. Glouberman, S.; Gemar, M.; Campsie, P.; Miller, G.; Armstrong, J.; Newman, C.; Siotis, A.; Groff, P. A framework for improving health in cities: A discussion paper. J. Urban Health 2006, 83, 325–338.
30. Bentley, M. Healthy Cities, local environmental action and climate change. Health Promot. Int. 2007, 22, 246–253.
31. English, P.B.; Sinclair, A.H.; Ross, Z.; Anderson, H.; Boothe, V.; Davis, C.; Ebi, K.; Kagey, B.; Malecki, K.; Shultz, R.; et al. Environmental health indicators of climate change for the United States: Findings from the state environmental health indicator collaborative. Environ. Health Perspect. 2009, 117, 1673–1681.

32. Centers for Disease Control and Prevention (CDC). National Public Health Performance Standards Program (NPHPSP); CDC: Atlanta, GA, USA, 2010. Available online: www.cdc.gov/nphpsp/ (accessed on 27 July 2011).

33. National Research Council (NRC). State of the USA Health Indicators: Letter Report. Committee on the State of the USA Health Indicators; National Academies Press: Washington, DC, USA, 2008. Available online: http://www.nap.edu/catalog/12534.html (accessed on 27 July 2011).

34. Balbus, J.M.; Malina, C. Identifying vulnerable subpopulations for climate change health effects in the United States. J. Occup. Environ. Med. 2009, 51, 33–37.

35. Galea, S.; Ahern, J.; Karpati, A. A model of underlying socioeconomic vulnerability in human populations: Evidence from variability in population health and implications for public health. Soc. Sci. Med. 2005, 60, 2417–2430.

36. Merzel, C.; D'Afflitti, J. Reconsidering community based health promotion: Promise, performance, potential. Am. J. Public Health 2003, 93, 557–574.

37. Semenza, J.C.; Maty, S. Improving the Macrosocial Environment to Improve Health: A Framework for Intervention. In Macrosocial Determinants of Health; Galea, S., Ed.; Springer Media Publishing: New York, NY, USA, 2007; pp. 443–462.

38. Semenza, J.C. Case Studies: Improving the Macrosocial Environment. In Macrosocial Determinants of Health; Galea, S., Ed.; Springer Media Publishing: New York, NY, USA, 2007; pp. 463–484.

39. Ebi, K. Climate change and health risks: Assessing and responding to them through 'adaptive management'. Health Aff. (Millwood) 2011, 30, 924–930.

# CHAPTER 10

# Building Resilience against Climate Effects: A Novel Framework to Facilitate Climate Readiness in Public Health Agencies

GINO D. MARINUCCI, GEORGE LUBER, CHRISTOPHER K. UEJIO, SHUBHAYU SAHA, AND JEREMY J. HESS

## 10.1 INTRODUCTION

Mounting evidence, assembled in numerous national and international-level scientific assessments, strongly indicates that climate change will have broad and significant impacts on infrastructure and a wide range of sectors, including agriculture, transportation, water, and energy management [1,2]. At the nexus of these impacts lie the societal consequences of climate change, well illustrated by the range of challenges to public health. Climate change is likely to have broad public health impacts, from the direct impacts of weather extremes to shifting geographies of infectious diseases and the potential for destabilization of critical societal support

*Building Resilience against Climate Effects—A Novel Framework to Facilitate Climate Readiness in Public Health Agencies.* © *Marinucci GD, Luber G, Uejio CK, Saha S, and Hess JJ.* International Journal of Environmental Research and Public Health *11,6 (2014). doi:10.3390/ijerph110606433.* *Licensed under Creative Commons Attribution 3.0 Unported License, http://creativecommons.org/ licenses/by/3.0/.*

systems such as food, energy and transportation. This has led the Director General of the World Health Organization to declare that climate change will be the defining issue for public health in the 21st century [3].

The public health community has identified several potential constraints and barriers to public health adaptation to climate change. These constraints and barriers are not unlike those facing several other sectors [4]. The barriers in public health arise from various sources and include uncertainty about future socioeconomic and climatic conditions as well as a range of financial, institutional, and cognitive limits within public health institutions that can constrain recognition of and action on climate change [5]. To address these barriers, Huang and colleagues have argued that public health needs to put a high priority on research clarifying the potential health impacts of climate change, including scenario-based projections of climate change health impacts; to clarify health co-benefits of potential mitigation strategies; and to evaluate the cost-effectiveness of potential adaptation options [5]. While some of these activities build on established programs and conventional strengths in public health [6], others—particularly scenario-based projections of climate change health impacts—will require the public health community to develop new skills and methodologies, forge new tools, and build new partnerships.

The United States Centers for Disease Control and Prevention (CDC), through the work of its Climate and Health Program, is one of the lead entities supporting state, tribal, local and territorial public health agencies to adapt to climate change in the U.S. [7]. One difficulty that has been identified in readying public health agencies is that climate change is a relatively novel concern and frontline actors in state and local agencies feel unprepared to address the challenges posed by climate change as a result of incomplete knowledge, inadequate staffing, and insufficient training in activities required to facilitate climate change adaptation [8]. Linked with these challenges is the diversity of exposures and health impacts affected by climate change, which includes direct and indirect exposures and associated health impacts [9] that will unfold via place-specific pathways [10]. In many cases our understanding of the systems in which these impacts will unfold is incomplete.

To build climate change adaptation capacity in the public health community, CDC has devised a comprehensive framework for developing local climate change adaptation plans. The Building Resilience Against Climate Effects (BRACE) framework is an iterative approach to adaptively manage the health effects of climate change. The adaptive management approach has shown promise in other sectors, such as the water and natural resources management sectors [11,12] and has been proposed as a potentially useful strategy in public health [11,12,13]. Moreover, BRACE is designed to clarify and assess the most concerning public health risks in a given region, acknowledging the place specificity of many emerging climate change health threats [10]. The CDC's Climate and Health Program is elaborating BRACE through its implementation in the Climate-Ready States and Cities Initiative cooperative grant program. Through this grant program BRACE is being deployed in 18 public health agencies.

Adaptive management is an iterative, learning-based approach to the design, implementation, and evaluation of interventions in complex, changing systems [14]. Adaptive management explicitly acknowledges that complex systems are incompletely understood, that management interventions can affect system behavior in unexpected ways, and that management strategies need to be regularly updated as system managers and stakeholders learn through interactions with the system and each other [14,15]. Systems well suited to an adaptive management approach are typically incompletely understood and sometimes exhibit unexpected attributes in response to management efforts; ecosystems are an oft-cited example [16]. Adaptive management principles posit that the key to managing these complex, non-linear systems is a learning-based strategy that uses models and emphasizes the need to periodically gather information about their behavior, particularly in response to management actions and shifting stressors over time.

As set forth by the National Research Council in 2004, the adaptive management approach is best applied in settings that exhibit six key elements:

1. Management objectives that are regularly revisited and revised;
2. A model of the system(s) being managed;
3. A range of management choices;
4. Monitoring and evaluation of outcomes;
5. Mechanisms for incorporating learning into future decisions; and

6.  A collaborative structure for stakeholder participation and learning [15].

BRACE incorporates adaptive management principles in its recognition that the public health impacts of climate change are complex and that the many systems involved are incompletely understood. BRACE relies explicitly on the use of modeling to project climate change health impacts and regular, iterative reassessment of risks and management priorities as well as integration of new knowledge about how the systems under management are likely to respond to management interventions. In recognition of other barriers and constraints to public health adaptation to climate change identified in the literature, BRACE also requires vulnerability assessment and evidence-based evaluation of potential intervention options. Altogether, these principles are aimed at including recent scholarship on climate change adaptation more generally as well as scholarship on the social determinants of health and integration of learning in public health systems, policy development, and planning [17,18,19,20,21,22,23].

We present the five sequential steps of BRACE and explain how the steps align with the adaptive management principles mentioned above.

## 10.2 BUILDING RESILIENCE AGAINST CLIMATE EFFECTS (BRACE)

BRACE provides for the systematic use of climate projections to inform public health adaptation efforts at a local or regional level [24]. The framework incorporates an assessment of climate change impacts, a vulnerability assessment, the modeling of projected health impacts, an evidence-based evaluation of intervention options, a strategy for implementing interventions, and systematic evaluation of all activities in an iterative framework. Broad stakeholder engagement, adaptive management, and a long range planning frame are key elements.

"Adaptation" can be defined as adjustments in ecological, social, or economic systems in response to actual or expected climatic stimuli and their effects or impacts. It refers to changes in processes, practices, and structures to moderate potential damages or to exploit beneficial opportunities associated with climate change [25,26]. In the context of BRACE, public health adaptation can be defined as any short- or long-term strategies that can reduce adverse health impacts or enhance resilience in response to observed or expected changes in climate and associated extremes [5]. Interventions are an important component of adaptation for climate change. An intervention is defined as a set of actions with a coherent objective to bring about change or produce identifiable outcomes. These actions may include policy, regulatory initiatives, single strategy projects or multi-component programmes. Public health interventions are intended to promote or protect health or prevent ill health in communities or populations [27]. In this paper, we regard interventions as actions that will bring about a change or produce an outcome that reduces adverse health impacts. Adaptations other than interventions can be considered strategies that prepare more resilient systems by focusing on building capacity and key capabilities in the public health agency and its partners.

There are five sequential steps in BRACE. Each step aligns with key elements of adaptive management. See Table 1.

**TABLE 1:** There are five sequential steps in BRACE.

| Step No. | BRACE Step Title | Description of Functions | Corresponding Adaptive Management Element |
|---|---|---|---|
| Step 1 | Anticipating Climate Impacts and Assessing Vulnerabilities | Identify the scope of climate impacts, associated potential health outcomes, and populations and locations vulnerable to these health impacts. | 1, 2, 4, 5, 6 |
| Step 2 | Projecting the Disease Burden | Estimate or quantify the additional burden of health outcomes due to climate change. | 1, 2, 4, 5 |
| Step 3 | Assessing Public Health Interventions | Identify the most suitable health interventions for the health impacts of greatest concern. | 1, 3, 4, 5, 6 |
| Step 4 | Developing and Implementing a Climate and Health Adaptation Plan | Develop a written plan that is regularly updated. Disseminate and oversee the implementation of the plan. | 1, 4, 6 |
| Step 5 | Evaluating Impact and Improving Quality of Activities | Evaluate the process. Determine the value of information attained and activities undertaken. | 1, 3, 4, 5, 6 |

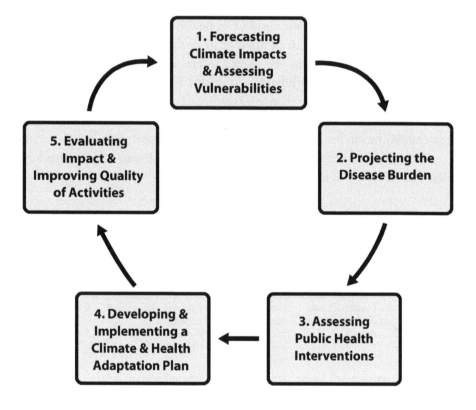

**FIGURE 1:** The iterative nature of BRACE.

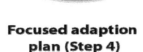

Comprehensive synopsis of climate
and health threats for
the jurisdiction (Step 1)

Projected future disease
prevalence (Step 2)

Intervention
options (Step 3)

Focused adaption
plan (Step 4)

**FIGURE 2:** Prioritization of climate and health impacts and suitable interventions.

## 10.2.1 THE FIVE STEPS OF BRACE

In line with a key element of adaptive management, BRACE is iterative and enables the incorporation of learning into future decisions. The novel consideration of climate change as a public health concern purports the need for flexibility to insert new or revised information, and improved analyses to assure better decisions can be made. BRACE allows for the inclusion of such information and the re-visit of the Steps to improve decision making and prioritizations throughout the cycle. See Figure 1.

BRACE enables systematic prioritization of adaptations for resource challenged public health agencies. From the overarching information detailed in Step 1, the hazards of greatest concern can be identified for projection. Step 2 enables further narrowing of scope by providing future disease burden estimates that may help a public health agency choose what issues are of the highest priority for taking action. Additionally, Step 3 allows for decisions about the most effective and locally suitable means for taking action, further prioritizing interventions. These steps culminate in the development and implementation of a locally tailored adaptation plan for public health. See Figure 2.

We describe the steps of BRACE that can be implemented by public health agencies of various sizes and locations and the key considerations for agencies embarking on becoming climate-ready.

## 10.2.2 STEP 1: ANTICIPATING CLIMATE IMPACTS AND ASSESSING VULNERABILITIES

The first step of BRACE involves two inter-related tasks: (1) working with weather, climate variability, and change data sources to identify climate sensitive health outcomes; and (2) identifying vulnerable populations. More specifically, the first task focuses on finding data sources, relating weather to health outcomes and identifying the range of health outcomes that may be affected by climate change and variability within the jurisdiction. Our description includes a summary of key features of climate change projections. By the conclusion of this task a public health agency will have the information to prepare a climate and health profile report

(CHPR) that compiles the list of health outcomes of concern for a jurisdiction and details the climate-health exposure pathways.

Then after familiarizing itself with the range of likely climate changes and public health impacts, a public health agency will characterize vulnerability within a jurisdiction. For this second task, we detail multiple methods for describing vulnerability. The CHPR and vulnerability assessment outputs generated in Step 1 serve the purpose of summarizing the range of climate change impacts of concern in the geographic area for policy makers and key stakeholders, and will present information to inspire further investigation and analysis by the public health agency or other interested entities.

## 10.2.2.1 TASK #1: WORKING WITH WEATHER, CLIMATE VARIABILITY, AND CLIMATE CHANGE DATA SOURCES TO IDENTIFY CLIMATE SENSITIVE HEALTH OUTCOMES

The public health community will use these data sources to: (a) understand projected climatic changes; (b) identify local weather/climate variability and human health relationships; and (c) obtain climate change projections for quantifying the future disease burden (BRACE Step #2). This can be challenging as public health practitioners are not routinely trained to find, manipulate and interpret this type of information. Working with weather, climate variability, and change information requires specialized technical knowledge. Public health agencies are encouraged to consult with local or regional climatologists who are primary stakeholders for this first task of BRACE.

Organizations skilled in translating climate information (e.g., the National Oceanic and Atmospheric Agency's Regional Integrated Science and Assessment Centers, State Climatologists) can also help a public health agency to acquire data if pursuing its own analyses (see task #2). Globally, there is significant variance in the development of comparable weather or climate service institutions. The WMO highlights this variance, while demonstrating the importance of such services when available. For example, the WMO highlighted examples of how climate information was improving decision making in Haiti, Mozambique, Fiji, Australia, and China [28]. In addition to these institutions, multiple websites aggregate and direct users to reliable sources of global weather information

[29,30,31]. Atmospheric scientists and climatologists have also shared "best practices" for translating and using climate variability and change information to improve decision making [32]. Similar to adaptive management, important practices include mutual education and trust building, joint-knowledge production, sustained collaboration, and sensitivity to organizational and cultural structures that may limit using climate information [33,34,35].

We provide background on climate change projections which are produced by multiple global climate models (GCM) that are driven by future greenhouse gas emissions. GCM are complex computerized simulations of the Earth's surface, oceans, cryosphere and atmosphere. Models project conditions over relatively large areas (~50–200 km). The clearinghouse for global projections is The World Climate Research Programme Coupled Modeled Intercomparison Project 5 (CMIP5) [36]. Over twenty global climate modeling consortiums share their GCM projections through the data portal. Projecting climate 30 or more years into the future is difficult because uncertainty is created by natural climate variability, incompletely understanding climate feedbacks and climate sensitivities, and societal actions and behaviors [37,38].

Projections are often used to quantify a range of uncertainty surrounding future conditions. Projections are derived from multiple GCM and involve representative concentration pathways (RCP) which cover a broad range of future greenhouse gas (GHG) emissions levels [39,40]. The upper end of the RCP range corresponds to rapid GHG emissions growth (RCP 8.5) while the lower RCP presumes aggressive GHG limits will be enacted (RCP 2.6). The middle pathways suggest that GHG stabilize at different levels (RCP 4.5, 6.0) by the end of the century.

Global data may not be detailed enough for some public health applications. Downscaling translates global climate data into locally relevant climate information. The North American Regional Climate Change Assessment Program and the United States Geological Survey provide state-of-the-art, dynamically downscaled, regional (~50 km) U.S. climate information [41,42]. Comparable initiatives exist for the Mediterranean, South America, East Asia, and Europe (e.g., [43]). We refer readers to in-depth discussions of climate models and downscaling techniques [44,45].

Global climate is projected over different multi-decadal periods from the year 2006 to 2300. Multiple GCM projections are made for each period and RCP since the modelling process is sensitive to initial conditions. Therefore, climate impact studies commonly work with the ensemble or average of GCM runs (1–9 runs) from each model. The ensemble is frequently more accurate than any individual projection [46]. The averaging process tends to cancel random GCM run errors which create an ensemble with the most agreement between runs. Researchers commonly employ two strategies to select GCM projections to use in impact studies. The first strategy presumes GCM have similar levels of skill to reproduce key climatic processes and selects all GCM or a random sample (~10 models) (e.g., [47]). The second strategy selects a subset of GCM that best simulate observed climatic processes. Each global climate modelling group projects future conditions and retrospectively simulates recent climatic conditions (e.g., 1986–2005). For example, researchers concerned with monsoonal precipitation may select GCM with the strongest correspondence between the seasonality and geographic pattern of retrospective simulations and observations (e.g., [48]).

International and national climate change and human health assessments provide an overview of the large range of health outcomes sensitive to climate [9,49,50]. Many climate and health relationships are location-specific. A classic example is an extreme heat event in Oslo, Norway may have the same physical characteristics as an average summer day in Mediterranean Montpellier, France. On average, residents of Montpellier may be more acclimatized (physiological, behavioral, built environment) to extreme heat than people in Oslo [51]. Public health agencies would, therefore, have to develop evidence for local climate and health relationships. Developing these relationships occurs in Step 2 of BRACE by using existing surveillance systems that collect morbidity (e.g., syndromic surveillance) and mortality information. The CDC National Environmental Public Health Tracking Network is an example of a complementary resource that integrates health, exposure, and hazard information into a web-based tool to analyze health impacts associated with environmental exposures that can be used to determine local climate and health relationships. [52,53]. Locally customized efforts are underway in other parts of the world, for example tailored heat response plans are being developed and implemented in India [54] and similarly customized early heat warning systems have been found to be effective in China and Hong Kong [55].

## 10.2.2.2 TASK #2 IDENTIFYING VULNERABLE POPULATIONS

A vulnerability assessment typically evaluates "the degree to which a system is susceptible to injury, damage, or harm" [56]. Key vulnerability concepts are exposure, sensitivity, and adaptive capacity. Exposure refers to the magnitude, frequency, and duration of an environmental exposure or disease risk. Sensitivity is the ability to withstand the exposure and its aftermath. Adaptive capacity refers to the broad range of responses and adjustments to the potential impact.

There are several methods used to conduct a vulnerability assessment. Vulnerability assessments can take on both qualitative and quantitative components. Quantitative studies tend to assess magnitude and location, to efficiently analyze vulnerability on a large scale. Qualitative studies tend to provide insight of household and individual level pressures from societal factors that are difficult to assess through quantitative methods.

Two approaches to identify vulnerability using Geographic Information Systems include: overlay analyses of risk factors and spatial ecologic studies. Both study designs require locational information (e.g., latitude, longitude) where cases are potentially exposed to hazards. An overlay analysis combines multiple layers of risk factors to spatially define and assess potential vulnerabilities to climate change [57,58,59]. This approach can efficiently analyze vulnerability over large areas, which may be beneficial to populous areas. Publicly available sociodemographic data (e.g., U.S. American Community Survey, tax parcels) provide information on societal vulnerability. Many countries periodically collect comparable census information (e.g., Integrated Public Use Microdata Series) at geopolitical units that can be as small as the neighborhood level [60]. In the absence of existing information, carefully planned household surveys may also provide useful vulnerability assessment information [61,62]. Remotely sensed information (e.g., satellites, Radio Detection and Ranging (RADAR)) can provide neighborhood-level environmental exposure information such as outdoor heat exposure or extreme precipitation rates.

A spatial ecologic study can build upon overlay analysis [63,64,65]. The study determines the most important exposure and vulnerability characteristics associated with a health outcome from observed data. For example, after controlling for population size and other confounders, investigators in Philadelphia, USA, learned that increasing the proportion of vacant households in a local neighborhood by 10%, increased the odds of extreme heat mortality by 40% [64]. Further analysis can be undertaken to assess infrastructure systems that may compound risk, such as distance to hospitals and clinics or utility service areas.

In summary, Step 1 of BRACE draws from multiple elements of adaptive management. For example, management objectives derived from prioritizing climate-sensitive health outcomes and their corresponding planning time frames are chosen in consultation with stakeholders such as local public health agencies, elected officials, planners, academics, community members, and non-governmental organizations (Elements 1, 6). Such consultation can be formal, informal, and structured or unstructured, including meetings, focus groups, open discussion fora, and surveys, among other approaches. Partnering with climatologists will help public health agencies to interpret and work with climate and climate change information (Element 6). Models of climate-sensitive health outcomes can be used to consider both public health risk and societal vulnerability (Element 2).

## 10.2.3 STEP 2: PROJECTING THE DISEASE BURDEN

At the conclusion of Step 1, a public health agency has developed a profile of how the climate is changing, the likely effects on health, and the populations and systems most vulnerable to these changes. In Step 2, the agency examines shifting disease burdens more closely. Step 2 can be done qualitatively to yield a general impression of how climate change may affect the risk for certain outcomes, at least capturing general climatic trends and environmental exposures, population vulnerability, and expected human health impacts. However, a quantitative effort, the results of which can be used in comparative health assessments and cost-benefit analyses, is likely to be of greater use to a range of stakeholders if the relevant exposure pathways are adequately understood and if there is sufficient data to drive projections.

From a conceptual standpoint, climate change health impact projections have several major components, as noted in Figure 3.

While a detailed presentation of this process is outside this paper's scope, the literature on projection of climate change-related health impacts is growing rapidly and there are several different studies that detail relevant methods [66,67,68]. Regardless of the specifics of the chosen approach, the initial step is to define the health outcomes of interest (identified in Step 1 of BRACE) and the climate-health exposure pathway(s) to be assessed. The changing climate presents a novel type of public health challenge in which historic climate and health relationships need to be revisited. Formal quantitative risk assessment using anticipated future disease burden is an unfamiliar exercise for many local public health agencies [8].

There are multiple methods for analyzing and quantifying existing climate and health exposure pathways. We briefly discuss two of the most common statistical methods. Time series and case-cross over studies are similar methods of retrospectively associating weather and climate with morbidity and mortality [69]. These popular study designs have been applied to air pollution, extreme heat, and infectious diseases (e.g., [70,71,72,73]). In a time series, the proportion of cases that temporally coincided with weather events is potentially attributable to weather [74]. Similarly, a case-crossover study compares individual exposures (e.g., ambient pollen levels) before a case-defining event (e.g., an emergency department visit for an asthma exacerbation) to exposures at comparable periods (e.g., ambient pollen levels on the same day as the emergency department visit one week prior or hence) [75].

A population (time series) or individual (case-crossover) is compared against itself to implicitly control for time invariant confounders. Some exposure-response functions are likely to vary by location; if this is anticipated, developing location-specific response functions, if data allow, can help address this heterogeneity. Time series statistics can also be used to evaluate the efficacy of a public health intervention (Adaptive Management Element 4) [76,77].

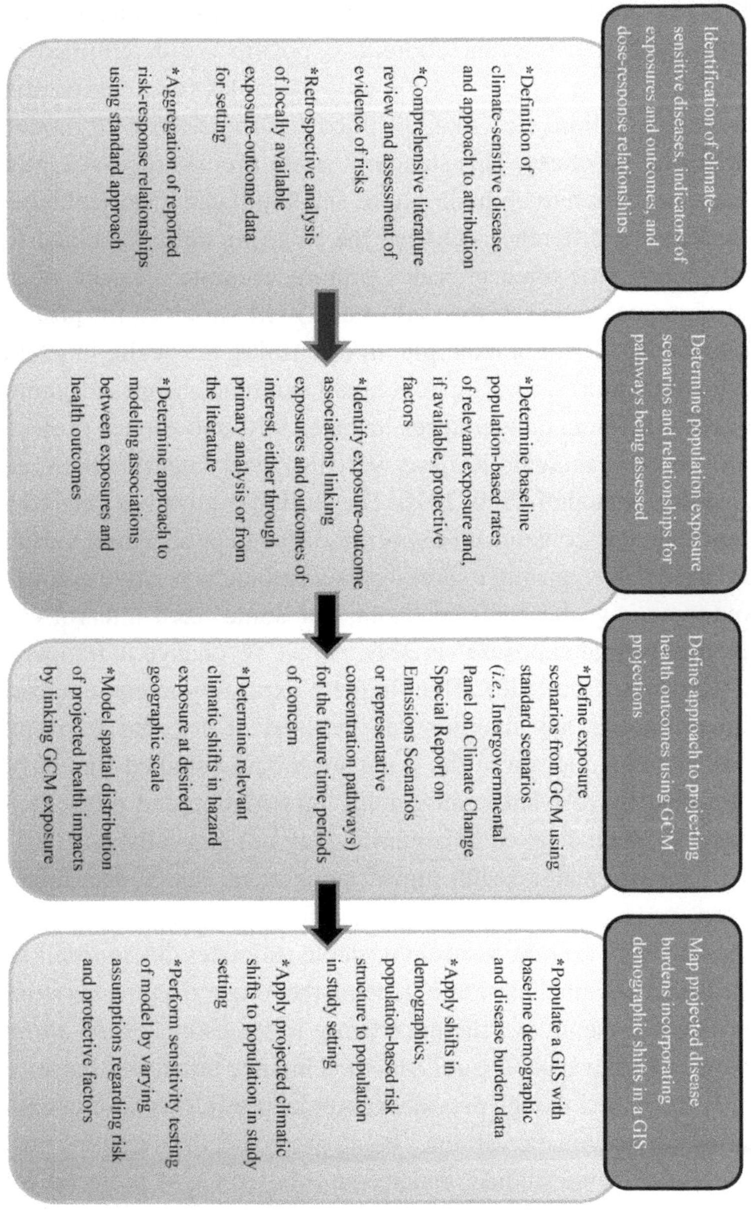

**FIGURE 3:** Major steps in projection of climate-associated disease burdens (adapted from [7] with permission from Springer®).

**Identification of climate-sensitive diseases, indicators of exposures and outcomes, and dose-response relationships**

*Definition of climate-sensitive disease and approach to attribution

*Comprehensive literature review and assessment of evidence of risks

*Retrospective analysis of locally available exposure-outcome data for setting

*Aggregation of reported risk-response relationships using standard approach

**Determine population exposure scenarios and relationships for pathways being assessed**

*Determine baseline population-based rates of relevant exposure and, if available, protective factors

*Identify exposure-outcome associations linking exposures and outcomes of interest, either through primary analysis or from the literature

*Determine approach to modelling associations between exposures and health outcomes

**Define approach to projecting health outcomes using GCM projections**

*Define exposure scenarios from GCM using standard scenarios (i.e., Intergovernmental Panel on Climate Change Special Report on Emissions Scenarios or representative concentration pathway's) for the future time periods of concern

*Determine relevant climatic shifts in hazard exposure at desired geographic scale

*Model spatial distribution of projected health impacts by linking GCM exposure

**Map projected disease burdens incorporating demographic shifts in a GIS**

*Populate a GIS with baseline demographic and disease burden data

*Apply shifts in demographics, population-based risk structure to population in study setting

*Apply projected climatic shifts to population in study setting

*Perform sensitivity testing of model by varying assumptions regarding risk and protective factors

Several different methods have been used to project disease impacts. While there is no firm consensus regarding the most appropriate approach [78], the most commonly applied technique is the Delta Method, in which changes in the relevant climatic exposure are determined by comparing each GCM's projected climatic variables (e.g., temperature, humidity, precipitation, etc.) in a specified future period with model simulated historical baselines. This standardization process removes model-specific biases common to both the future and simulated historical baseline. Next, the relative difference between the two (the delta) is added to historical baselines. This often provides a more accurate measure of shifting environmental exposure than adjusted GCM variables for projecting future disease burdens. For example, in examining the health impact of increasing temperatures, investigators might examine changes in summertime average maximum temperatures in June, July, and August projected to 2035 and compare these projections with historical data from the specified baseline time period of 1980–2010. The shifted exposure is entered into a function that also contains an exposure-outcome association variable, often a relative risk (typically expressed as a change in relative risk of a specified outcome, e.g., heat-related death, per some fixed interval change in an environmental exposure variable, e.g., 1 °C change in temperature above a particular threshold). The simplest exposure-outcome associations are linear, though any function can be used. If possible, exposure-outcome associations relevant to the population being studied should be used and stratified by relevant demographic information and other factors, but in practice, such detailed information may not be available.

Climate change health impact projections can be data intensive. However, many uncertainties arise, and efforts to increase the precision of projected outcomes may result in unstable estimates due to small sample sizes and other potential biases. In general, the longer the time horizon used in the projection, the greater the uncertainty in the estimates. At a minimum, the data needed to apply the delta method include baseline disease rates (preferably incidence, though prevalence can be used), exposure-outcome associations for the climatic variables being projected, demographic projections for the region being studied, and scenario-based GCM projections of shifts in climatic variables in the study region for the study period. Considerations regarding data sources for projections in Step 2 are listed in Table 2.

**TABLE 2:** Common data sources used in climate change health impact projections (adapted from Hess et al. [7] with permission from Springer®).

| Category of Data Required | Common Data Sources |
|---|---|
| Baseline rates of disease | Ongoing public health surveillance; published and unpublished regional and national datasets (e.g., National Hospital Ambulatory Medical Care Survey; Healthcare Cost and Utilization Project; Nationwide Emergency Department Sample; Behavioral Risk Factor Surveillance System). |
| Exposure-outcome associations | Published literature; retrospective analysis of local health outcome datasets merged with local weather and climate data from the National Climatic Data Center or another source; CDC National Environmental Public Health Tracking Network. |
| Demographic projections | Demographic projections available for the country as a whole from the United States Census; available for individual states via the Federal-state Cooperative for Population Projections. |
| GCM projections | There are a number of climate models worldwide and certain outputs have been made publicly available; one commonly used source is the Coupled Model Intercomparison Project (CMIP), which issues ensemble model runs for various scenarios (e.g., CMIP3, CMIP5) that are available for download [40,79]; these projections use standard scenarios (e.g., Intergovernmental Panel on Climate Change Special Report on Emissions Scenarios and RCPs) [80]. |

A public health agency should consider whether adaptation should be included in the model. Adaptation activities have the potential to reduce the adverse impacts of climate change [5]. Reasons for including adaptation include general plausibility and methodological consistency specifically if other factors, e.g., demographics shifts that may affect vulnerability are included in the model, in which case it would be reasonable to include adaptation as well. Reasons for leaving out adaptation include a lack of consensus regarding how to model adaptation or an approach in which other factors affecting vulnerability are explicitly left out. Depending on the length of study period (i.e., how far into the future health impacts are projected), projections of likely adaptations—active and passive, planned and unplanned—will be more or less important, as more adaptations will presumably be employed with longer timeframes. There are many different adaptations, some passive (e.g., physiologic adaptation to increasing ambient temperatures) and some active (e.g., usage of mechanical air conditioning) that might be considered.

Including adaptations presumably reduces the likelihood that projected disease burdens will be systematic overestimates, particularly in the long term, and several different methods have been used, but there are no agreed-upon approaches. In some cases, physiologic adaptation to the exposure of concern has been incorporated into exposure-outcome response functions [67]. This approach is viable when there is physiological adaptation, as is the case with ambient temperature changes. In other cases, adaptation has been accounted for by systematically adjusting or discounting estimates of future impacts [81]; this approach is more generic and can be used as a proxy for many different types of adaptation or a mixed combination of adaptation strategies.

While projecting disease burden can be time- and data-intensive, the investment can potentially pay dividends going forward as the models can be used to guide several different types of decisions over time and to engage with various stakeholders to help prioritize risk management decisions. The models can also be further developed and modified as additional information becomes available and thereby feed into adaptive management processes [82].

### 10.2.4 STEP 3: ASSESSING PUBLIC HEALTH INTERVENTIONS

The focus of Step 3 of BRACE is identifying the most suitable interventions to adapt to the climate change related health threats identified as of greatest concern in Steps 1 and 2. For various clinical and public health interventions, the evidence-based public health (EBPH) approach provides practitioners an opportunity to assess the efficacy of alternative interventions [83]. While the EBPH literature on public health interventions in response to climate change is meager, a recent review lists a wide range of such interventions [84] and outlines the relevant evidence.

The general EBPH approach entails the following steps: (i) assessment of the problem; (ii) a systematic review of the public health literature to identify relevant interventions; and (iii) assessment of the efficacy of interventions [83,85]. In the case of climate change, the problem assessment step includes both assessment of the shifting exposures resulting from climate change and the likely health impacts [86]. Step 3 of BRACE is focused, in particular, on the latter parts of the EBPH process, in which intervention efficacy is examined closely.

While the literature on some potential public health impacts of climate change (e.g., problem assessment) is substantial, there is relatively little information published on specific adaptations and interventions that may avoid or limit these projected impacts. For instance, a recent structured review of population-level interventions to reduce the impacts of extreme heat identified only 14 studies, all of which were cross-sectional or retrospective, and the authors were unable to generate a specific estimate of impacts [87]. As instances of agencies designing and implementing these health interventions across different jurisdictions increase, established guidelines like the Preferred Reporting Items for Systematic Reviews and Meta-analyses [88] could be used to evaluate and compare intervention efficacy.

Even when a range of evidence-based health interventions are available, budget constraints may necessitate comparisons based on cost to help public health agencies prioritize and select across potential alternatives. The comparative cost-effectiveness paradigm [89] provides public health agencies an approach to choose among available intervention and adaptation strategies in this context. By adapting the definition of comparative effectiveness in clinical research developed by the Institute of Medicine [90], this approach requires systematic comparison of available estimates of costs and benefits associated with prevention, diagnosis, treatment, and monitoring of alternative strategies designed to reduce the adverse health impacts from changes in climate-sensitive exposures. Established protocols on conducting comparative effectiveness research using observational data [91] and cost-effectiveness analysis [92] can be adapted to evaluate and identify intervention strategies. While higher levels of evidence are particularly useful for more costly interventions, other forms may be very important in guiding day-to-day decisions that many public health officials encounter in the course of their activities. Some locales may decide it is more appropriate for them to collect their own evidence to guide interventions. Public health practitioners also frequently use anecdotal evidence conveyed through informal professional networks in making ad-hoc decisions when infrequently studied issues arise, such as strategies for promoting the use of cooling centers and making decisions about when to issue heat-health warnings.

Evidence may also not be available for certain potential risks, particularly those associated with cascading system failures like electrical blackouts or sewage treatment failures after extreme precipitation events. In such cases, public health officials may need to access literature outside of public health to identify strategies for promoting resilience across a range of linked systems upon which public health relies.

Overall, while systematic review of the literature and identification of efficacious interventions is of paramount importance, it is also clear that other, less robust forms of evidence will also enter into deliberations regarding the interventions to pursue. As the field matures and various interventions are implemented, public health practitioners may consider prioritizing the reporting of these interventions and their effects using relevant guidelines already in the literature.

In addition to assessing literature and cost-effectiveness for interventions, a public health agency may also consider assessing the suitability of interventions to their political, cultural, and logistical environment. In this instance, a public health agency may consult with a range of stakeholders that will be affected by or play a role in proposed interventions. For example, considering adaptations to extreme heat may require coordination inside public health agencies, with other health partners, local weather forecast offices, non-government organizations and community groups to establish critical temperature thresholds, consider the dissemination of information to vulnerable populations and identify locally appropriate response measures such as the activation of cooling shelters or transportation of immobile seniors to protective environments. Consultation might consider the resources needed and available for the considered interventions, the skill sets and technological assets available or needed, cultural and political palatability and the opportunity cost of taking action. Assessing these elements can provide insight as to the likely acceptance of the action and the likely barriers that will be faced.

Both assessments of intervention efficacy and suitability for the specified setting will feed into the process of identifying interventions that are most likely to be suitable for the jurisdiction. The combined assessments enable the ranking of a set of focused actions that will form the core elements of a locally specific and manageable adaptation plan for public health. Undertaking Step 3 of BRACE employs several elements of adaptive management. This step studies and considers a range of management choices (Element 3) and in its most collaborative form enables extensive opportunities for stakeholder participation and learning (Element 6).

## 10.2.5 STEP 4: DEVELOPING AND IMPLEMENTING A CLIMATE AND HEALTH ADAPTATION PLAN

The focus of Step 4 of BRACE is synthesizing information generated in the prior steps into a focused climate and health adaptation plan. BRACE emphasizes the need for a unified adaptation plan for the public health sector to foster collaboration across disciplines and interest groups and to align efforts to a common objective of protecting and promoting health. Creating such a plan is not a substitute for integrating public health considerations into broad climate change plans in other sectors. Rather, it should complement and add greater specificity to public health considerations that will be ideally included within a cross-sectoral, jurisdiction wide plan for climate change action.

With respect to adaptation, in practice, stakeholders may use disparate interpretations that reflect different underlying goals and priorities [93]. Explicitly recognizing these differences is important for reconciling goals and developing evaluation metrics for adaptation planning [94]. Frameworks, heuristics, and "best practices" from the climate and global change adaptation and sustainability science literature have been incorporated into BRACE [20,95]. For example, stakeholder engagement, communication, iterative learning, and adaptive management are key principles of both "successful" adaptation frameworks [25,94] and BRACE, and are reflected in BRACE guidance for climate and health adaptation plans.

A climate and health adaptation plan might aim to coordinate, highlight, and, potentially instigate a series of activities aimed at preventing, or at least reducing, the anticipated impacts of climate change in the area. Activities in the plan may be comprised of new, enhanced, or established programs and activities and the plan may outline or reinforce how these activities will be implemented and their success measured. The sponsoring public health agency may also consider including a means to engage with other key sectors that have responsibility for policies, programs, or oversight that will ultimately affect public health outcomes. For example, developing memorandums of understanding that foster greater exchange of information, skills and resources between a public health agency and agencies responsible for housing, planning, transportation and water quality, might spur earlier input into development plans that can further optimize public health benefits or mitigate unintended harm. A public health agency might consider a means for the adaptation plan to be periodically updated, and ensuring public accessibility of the plan.

In preparing an adaptation plan, a public health agency should aim to be comprehensive, considering options for action that cut across all of the essential public health functions from surveillance to regulation to outreach and education [6]. However, a plan must also be balanced with the feasibility considerations detailed in Step 3 to increase the probability of plan uptake and implementation. A plan should identify how stakeholders can integrate adaptations into their existing programs and should detail how activities will be evaluated. The scope of the plan should be intersectoral while maintaining a focus on public health outcomes. The planning horizon for the activities contained within the plan should span several years.

In preparing a climate and health adaptation plan, a public health agency can adapt the key elements and structure found in many jurisdiction wide, multisectoral climate change adaptation plans. A typical Climate Change Adaptation Plan has the following elements:

- Community profile which includes background information
- Most appropriate regional/municipal climate change scenario
- Scoped local climate change impacts
- Prioritized consequences/prospects of risks and opportunities
- Maps showing priorities
- Adaptation planning principles
- Table of recommended adaptation policies and actions indicating priority, lead responsibility and fit with existing program (if applicable)
- Action plan for tasks to be accomplished in the community
- Community engagement process
- List of key stakeholders
- Inventory of risks and opportunities
- Inventory of consequences and prospects
- Gap analysis of programs useful for adaptation actions [96]

Public health agencies should anticipate that climate and health adaptation plans will be used as internal and external communication documents. Therefore plans should clearly outline the resources required to undertake the series of activities, detail how established programs may need to be modified to cater for changing risk factors and environmental conditions, and identify the parties responsible for executing the activities. As discussed earlier, some responsibilities may lie in partnerships with other agencies. The role of these agencies, the mechanism for their engagement and the means for coordinating should be included in the plan, and the nature of the collaboration must be well described. The public health agency should include clear language expressing its commitment to communicate its findings and updates to stakeholders throughout the course of the plans implementation. For stakeholders outside of the public health agency, the plan will provide a vision for protecting health against the effects of climate change in the jurisdiction and can be used to educate prospective partners on ways they can contribute to the strategy. Finally, the plan should explicitly detail future review and periodic revisions of the plan to ensure that the projected health threat and adaptation strategies do, in fact, represent the most appropriate path forward for protecting the public's health.

To initiate implementation of the plan the public health agency must assure its comprehensive dissemination among a wide range of stakeholders that may play a role in implementing the stated activities.

The process of developing and implementing a climate and health adaptation plan crosses all of the elements of adaptive management. In particular, a plan is a means for defining the management objectives (Element 1) of the strategy, the prioritized interventions and adaptations (Element 3), measures for monitoring and evaluating progress (Element 4), quality improvement processes (Element 5) and the role of stakeholders (Element 6).

## 10.2.6 STEP 5: EVALUATING IMPACT AND IMPROVING QUALITY OF ACTIVITIES

Implementing BRACE is an iterative process. Climate change considerations are relatively new for the public health community, therefore, gathering information on the processes used to address the health effects of climate change and the potential outcomes and impacts of those processes are critical. It is important to note that, while evaluation is positioned in Step 5 to better accommodate discussion and communication, in reality, monitoring and evaluation of processes, outcomes and key indicators are central considerations throughout the entire process. The value of explicitly making processes iterative is that management decisions can be revisited with new information, not only general knowledge about the threats being managed, but also information gleaned from experience since the process began. This ability to continually improve interventions is a fundamental adaptive management tenet.

At any point in the implementation of BRACE, process evaluation measures can help to validate methods employed and to reveal flaws in the plan. In addition to assessing the execution of key methods, process metrics can help to determine if the most appropriate stakeholders have been engaged and if the stakeholders' engagement added critical input. Evaluation can also identify the outcomes that resulted from the combined series of activities, for example, improved capacity to develop models, stronger partnerships with key stakeholders, identification of additional surveillance needs, and awareness of synergies across programs. Ultimately, the evaluation step can help to determine the impact of implementing BRACE by determining the extent to which the interventions improved public health outcomes. Evaluating impact is influenced by the quality of assessment performed in Steps 1 and 2 of BRACE. Rigorously evaluating public health intervention requires baseline climate and health relationships (Step 1) [97]. Similarly, adaptations can be evaluated against counterfactuals (Step 2) that estimate climate change attributable disease burden in the absence of adaptation.

While a comprehensive discussion of evaluation methodology is outside the scope of this paper, it is expected that monitoring and evaluation methods will be used in the implementation of BRACE to gauge progress. While each agency will have different evaluation resources, the agency should be able to answer some basic questions after its evaluation activities:

- Does the public health agency have a reasonable estimate of the health impacts of climate change in its jurisdiction?

- Has the process allowed the public health agency to prioritize health impacts of greatest concern and the most suitable interventions?
- Has the public health agency prepared an adaptation plan for the public health sector within the jurisdiction?
- Are climate change considerations accommodated in public health planning and implementation activities?
- Are public health considerations accommodated in climate change planning and implementation activities?
- Are indicators in place that will evaluate the interventions implemented as a result of utilizing BRACE?
- How can the process be improved in the next iteration?
- What are the agency's top learning priorities in the next iteration of BRACE?

The long-standing tradition of institutional learning from responses to novel threats places the public health community in a strong position to overcome many of the potential constraints and barriers to climate change adaptation identified in the introduction [5]. Public health agencies can accelerate the learning needed by programs to more effectively address the health impacts of climate change by employing rigorous monitoring and evaluation processes. To achieve this public health agencies may consider how they can sustain their commitment to learning, the need for new skills and consideration of longer planning horizons [5].

## 10.3 KEY CONSIDERATIONS FOR IMPLEMENTING BRACE

There are some key points to consider in the implementation of BRACE. First, eliciting stakeholder viewpoints and perspectives can add significant value to the overall process. While stakeholder engagement is critical in each step, BRACE is an adaptive management framework and therefore calls for periodic revisions to the stakeholder network in order to better align with the specific goals of each step. For example, in Steps 1 and 2, where the emphasis lies on the integration of climate change scenarios into the projection of health effects, a public health agency may gain significantly from engagement with the climate science community, as the expertise of generating and using climate projections may not exist within the agency. In Step 3, which focuses on the assessment and identification of appropriate public health actions and interventions, it would be appropriate to solicit input from the larger public health practitioner and health care provider communities and other non-health agencies that may share co-benefits.

A second point to consider while implementing BRACE is the temporal scope for assessing changing climatic conditions and for choosing adaptations or key interventions. Factors such as quality of data and certainty must be balanced against considerations of a public health agency's level of readiness and the future capacity needed to assure that the public health system will be able to cope with novel threats or significant increases in high risk exposures. A public health agency may anticipate needing to deploy certain interventions, which provide insight into key timeframes for planning and assessment. For example, a public health agency that anticipates that extreme heat events are of concern may target interventions linked to city planning. While the public health community operates on short-term horizons, the planning community deals with long term issues, mediating the use of space and shaping future development [98]. Typically, city or regionals plans are prepared with a 20–25 year planning horizon [99,100,101,102,103]. Planners need to have the understanding to integrate future health matters into their day-to-day considerations [98]. In order to provide scientifically credible inputs into these planning processes, public health agencies can benefit from modeling disease outcomes that may be influenced by environmental conditions in similar time periods.

A public health agency anticipating the effects of sea level rise on drinking water and wastewater infrastructure may consider a longer planning horizon. For example, the American Water Works Association estimates that drinking water and waste water pipes laid in post-World War II can be expected to last about 75 years. A public health agency that seeks to assess threats associated with sea level rise may need to focus on projected exposures and risk factors closer to a 75-year timeframe, to accommodate wastewater and drinking water infrastructure planning considerations [104].

The third key point relates to how a public health agency prioritizes the health impacts to be addressed and the interventions to be employed. Changing climatic conditions will directly or indirectly impact many health outcomes. This broad range of health outcomes, combined with limited experience with using and interpreting climate change projections, can be disconcerting and overwhelming for a public health manager considering a course of action to manage these threats [8,105]. Shrinking resources and competing demands on public health agencies further compound the challenges associated with taking action [106,107]. BRACE provides a system to manage the information provided in climate change projections, triage the health outcomes of concern, and prioritize the most suitable interventions for the jurisdiction.

## 10.4 SUMMARY AND CONCLUSIONS

Climate change is an evolving concern. For public health agencies, climate change presents a number of adaptation challenges, not least of which is the incompletely understood nature of the impacts and the ecological systems in which these impacts will unfold. Managing the public health risks associated with climate change requires an iterative framework consistent with the principles of adaptive management. BRACE incorporates the features of adaptive management into a stepwise process tailored for public health agencies. BRACE is currently in a proof-of-concept phase and is laid out here for dissemination beyond the locales in which it is being piloted. Following the steps laid out in BRACE should enable a public health agency to use the best available science to assess current and future climatic conditions and prioritize the health outcomes and interventions that are most important and suitable for the jurisdiction.

BRACE implementation thus serves as an opportunity for learning in public health. We support efforts of public health agencies implementing BRACE, to evaluate how the framework helps them negotiate various barriers and limits to climate change adaptation, and where it can be revised. BRACE incorporates "best practices" from the broader climate change adaptation literature while providing flexibility for public health agencies to develop innovative and rigorous adaptation case studies. Similarly, these agencies will evaluate how well adaptive management facilitates working with complex systems that they attempt to influence. BRACE pragmatically recognizes that public health and climate change are two of many considerations for making adaptation policy decisions.

Preliminary feedback from participants in CDC's Climate-Ready States and Cities Initiative, points to the complexity in accessing, interpreting and using climate model projections for empirically projecting the disease burden. This necessitates the use of sophisticated modeling expertise that may prove challenging for public health agencies. The level of specificity needed for disease projections will become more apparent as this information informs public health adaptations going forward. How successfully public health agencies navigate this challenge depends on the collaborations built with agencies involved in developing climate projections.

Acknowledging that BRACE may need to be refined as evidence regarding its implementation accumulates, we feel it is important to highlight that adopting BRACE serves as a ratification of public health's established commitment to evidence-based practice and institutional learning. Both will be paramount as public health wrestles with the significant new challenges that climate change presents. The implementation of BRACE across a variety of settings serves as an opportunity for evaluation of the framework's utility and, indirectly, of the utility of the principles of adaptive management upon which it is built. In this way, BRACE may serve to advance the science related to climate change adaptation not only within public health but more generally.

## REFERENCES

1.  Solomon, S. Climate Change 2007: Contribution of Working Group I to the Fourth Assessment Report of the Intergovernmental Panel on Climate Change; Cambridge University Press: Cambridge, UK, 2007.
2.  Karl, T.R.; Melillo, J.M.; Peterson, T.C. Global Climate Change Impacts in the United States; Cambridge University Press: Cambridge, UK, 2009.
3.  Chan, M. Climate Change and Health: Preparing for Unprecedented Challenges; David, E., Ed.; Fogarty International Center, National Institutes of Health: Bethesda, MD, USA, 2007; Volume 10.
4.  Moser, S.C.; Ekstrom, J.A. A framework to diagnose barriers to climate change adaptation. Proc. Natl. Acad. Sci. USA 2010, 107, 22026–22031.
5.  Huang, C.; Vaneckova, P.; Wang, X.; FitzGerald, G.; Guo, Y.; Tong, S. Constraints and barriers to public health adaptation to climate change. Amer. J. Prev. Med. 2011, 40, 183–190.
6.  Frumkin, H.; Hess, J.; Luber, G.; Malilay, J.; McGeehin, M. Climate change: The public health response. Amer. J. Public Health 2008, 98, 435–445.

7.  Hess, J.; Marinucci, G.; Schramm, P.; Manangan, A.; Luber, G. Management of climate change adaptation at the united states centers for disease control and prevention. In Climate Change and Global Public Health; Pinkerton, K., Rom, W., Eds.; Springer: New York, NY, USA, 2013.

8.  Maibach, E.W.; Chadwick, A.; McBride, D.; Chuk, M.; Ebi, K.L.; Balbus, J. Climate change and local public health in the united states: Preparedness, programs and perceptions of local public health department directors. PLoS One 2008, 3.

9.  Portier, C.; Thigpen-Tart, K.; Hess, J.; Luber, G.; Maslak, T.; Radtke, M.; Strickman, D.; Trtanj, J.; Carter, S.; Dilworth, C.; et al. A Human Health Perspective on Climate Change; Environmental Health Perspectives, National Institute of Environmental Health Sciences: Research Triangle Park, NC, USA, 2010.

10. Hess, J.; Malilay, J.; Parkinson, A.J. Climate change: The importance of place. Amer. J. Prev. Med. 2008, 35, 468–478.

11. Ebi, K. Adaptive Management to the Health Risks of Climate Change. In Climate Change Adaptation in Developed Nations; Ford, J., Berrang-Ford, L., Eds.; Springer: New York, NY, USA, 2011.

12. Ebi, K. Climate change and health risks: Assessing and responding to them through "adaptive management". Health Affir. 2011, 30, 924–930.

13. Hess, J.; McDowell, J.; Luber, G. Integrating climate change adaptation into public health practice: Using adaptive management to increase adaptive capacity and build resilience. Environ. Health Perspect. 2011, 120, 171–179.

14. Holling, C. Adaptive Environmental Assessment and Management; Wiley: New York, NY, USA, 1978.

15. National Research Council. Adaptive Management for Water Resources Project Planning; The National Academies Press: Washington, DC, USA, 2004.

16. Bormann, B.T.; Cunningham, P.G.; Brookes, M.H.; Manning, V.W.; Collopy, M.W. Adaptive Ecosystem Management in the Pacific Northwest; Pacific Northwest Research Station, Forest Service, U.S. Department of Agriculture: Portland, OR, USA, 2007.

17. Baker, E.A.; Metzler, M.M.; Galea, S. Addressing social determinants of health inequities: Learning from doing. Amer. J. Public Health 2005, 95, 553–555.

18. Berkhout, F.; Hertin, J.; Gann, D. Learning to adapt: Organisational adaptation to climate change impacts. Clim. Change 2006, 78, 135–156.

19. Füssel, H.M. Assessing adaptation to the health risks of climate change: What guidance can existing frameworks provide? Int. J. Environ. Health Res. 2008, 18, 37–63.

20. Füssel, H.M. Adaptation planning for climate change: Concepts, assessment approaches, and key lessons. Sustain. Sci. 2007, 2, 265–275.

21. Kelly, M.; Morgan, A.; Ellis, S.; Younger, T.; Huntley, J.; Swann, C. Evidence based public health: A review of the experience of the national institute of health and clinical excellence (nice) of developing public health guidance in england. Soc. Sci. Med. 2010, 71, 1056–1062.

22. Klinke, A.; Renn, O. A new approach to risk evaluation and management: Risk-based, precaution-based, and discourse-based strategies. Risk Anal. 2002, 22, 1071–1094.

23. Van Wave, T.W.; Scutchfield, F.D.; Honore, P.A. Recent advances in public health systems research in the United States. Annu. Rev. Public Health 2010, 31, 283–295.

24.  Marinucci, G.; Luber, G. Bracing for Impact: Preparing a Comprehensive Approach to Tackling Climate Change for Public Health Agencies. In Proceedings of the American Public Health Association Annual Conference 2011, Washington, DC, USA, 2 November 2011.

25.  Adger, N.W.; Arnell, N.W.; Tompkins, E.L. Successful adaptation to climate change across scales. Global Environ. Change 2005, 15, 77–86.

26.  United Nations Framework Convention on Climate Change Focus: Adaptation. Available online: https://unfccc.int/focus/adaptation/items/6999.php (accessed on 8 January 2014).

27.  Rychetnik, L.; Frommer, M.; Hawe, P.; Shiell, A. Criteria for evaluating evidence on public health interventions. J. Epidemiol. Community Health 2002, 56, 119–127.

28.  World Meteorological Organization. Climate Knowledge for Action: A Global Framework for Climate Services—Empowering the Most Vulnerable; World Meteorological Organization: Geneva, Switzerland, 2011.

29.  National Aeronautics and Space Administration. Global Change Master Directory. Available online: http://gcmd.nasa.gov/ (accessed on 12 August 2013).

30.  Centers for Disease Control & Prevention. Climate Change. Available online: http://www.cdc.gov/climateandhealth/technical_assistance.htm (accessed on 12 August 2013).

31.  National Center for Atmospheric Research. Climate Data Guide. Available online: https://climatedataguide.ucar.edu/ (accessed on 12 August 2013).

32.  Jacobs, K.; Garfin, G.; Lenart, M. More than just talk: Connecting science and decisionmaking. Environ.: Sci. Policy Sustain. Dev. 2005, 47, 6–21.

33.  Morss, R.E.; Wilhelmi, O.V.; Downton, M.W.; Gruntfest, E. Flood risk, uncertainty, and scientific information for decision making: Lessons from an interdisciplinary project. Bull. Amer. Meteorol. Soc. 2005, 86, 1593–1601.

34.  National Research Council. Completing the Forecast: Characterizing and Communicating Uncertainty for Better Decisions Using Weather and Climate Forecasts; The National Academies Press: Washington, DC, USA, 2006.

35.  Dilling, L.; Lemos, M.C. Creating usable science: Opportunities and constraints for climate knowledge use and their implications for science policy. Global Environ. Change 2011, 21, 680–689.

36.  World Climate Research Programme. Cmip5 Coupled Model Intercomparison Project. Available online: http://cmip-pcmdi.llnl.gov/cmip5/ (accessed on 12 August 2013).

37.  Hawkins, E.; Sutton, R. The potential to narrow uncertainty in regional climate predictions. Bull. Amer. Meteorol. Soc. 2009, 90, 1095–1107.

38.  O'Neill, B.C.; Dalton, M.; Fuchs, R.; Jiang, L.; Pachauri, S.; Zigova, K. Global demographic trends and future carbon emissions. Proc. Natl. Acad. Sci. USA 2010, 107, 17521–17526.

39.  Moss, R.H.; Edmonds, J.A.; Hibbard, K.A.; Manning, M.R.; Rose, S.K.; van Vuuren, D.P.; Carter, T.R.; Emori, S.; Kainuma, M.; Kram, T. The next generation of scenarios for climate change research and assessment. Nature 2010, 463, 747–756.

40.  Taylor, K.E.; Stouffer, R.J.; Meehl, G.A. An overview of cmip5 and the experiment design. Bull. Amer. Meteorol. Soc. 2012, 93, 485–498.

41. Mearns, L.O.; Arritt, R.; Biner, S.; Bukovsky, M.S.; McGinnis, S.; Sain, S.; Caya, D.; Correia, J., Jr.; Flory, D.; Gutowski, W. The north american regional climate change assessment program: Overview of phase i results. Bull. Amer. Meteorol. Soc. 2012, 93, 1337–1362.
42. Hostetler, S.W.; Alder, J.R.; Allan, A.M. Dynamically Downscaled Climate Simulations over North America: Methods, Evaluation, and Supporting Documentation for Users. Available online: http://pubs.usgs.gov/of/2011/1238/ (accessed on 16 June 2014).
43. Arritt, R.W.; Rummukainen, M. Challenges in regional-scale climate modeling. Bull. Amer. Meteorol. Soc. 2011, 92, 365–368.
44. Rummukainen, M. State-of-the-art with regional climate models. Clim. Change 2010, 1, 82–96.
45. Cooney, C.M. Downscaling climate models: Sharpening the focus on local-level changes. Environ. Health Perspect. 2012, 120, 22–28.
46. Pierce, D.W.; Barnett, T.P.; Santer, B.D.; Gleckler, P.J. Selecting global climate models for regional climate change studies. Proc. Natl. Acad. Sci. USA 2009, 106, 8441–8446.
47. Demaria, E.; Maurer, E.; Thrasher, B.; Vicuña, S.; Meza, F. Climate change impacts on an alpine watershed in chile: Do new model projections change the story? J. Hydrol. 2013, 502, 128–138.
48. Bukovsky, M.S.; Gochis, D.; Mearns, L.O. Towards assessing narccap regional climate model credibility for the north american monsoon: Current climate simulations. J. Climate 2013, 26, 8802–8826.
49. Parry, M.L. Climate Change 2007: Impacts, Adaptation and Vulnerability: Working Group II Contribution to the Fourth Assessment Report of the Ipcc Intergovernmental Panel on Climate Change; Cambridge University Press: Cambridge, UK, 2007; Volume 4.
50. Third National Climate Assessment Report; NCADAC: Washington, DC, USA, 2013.
51. Robinson, P.J. On the definition of a heat wave. J. Appl. Meteorol. 2001, 40, 762–775.
52. Centers for Disease Control and Prevention. Climate Change Indicators Available on the Environmental Public Health tracking Network. Available online: http://ephtracking.cdc.gov/showClimateChangeIndicators.action (accessed on 1 June 2012).
53. Centers for Disease Control and Prevention. National Environmental Public Health Tracking Network. Available online: http://ephtracking.cdc.gov/showHome.action (accessed on 1 May 2012).
54. Knowlton, K.; Kulkarni, S.; Azhar, G.S.; Mavalankar, D.; Jaiswal, A.; Connolly, M.; Nori-Sarma, A.; Rajiva, A.; Dutta, P.; Deol, B.; et al. Development and implementation of south asia's first heat-health action plan in ahmedabad (Gujarat, India). Int. J. Environ. Res. Public Health 2014, 11, 3473–3492.
55. Toloo, G.; FitzGerald, G.; Aitken, P.; Verrall, K.; Shilu, T. Are heat warning systems effective? Environ. Health 2013, 12.
56. Smit, B.; Wandel, J. Adaptation, adaptive capacity and vulnerability. Glob. Environ. Change 2006, 16, 282–292.

57. Cutter, S.L.; Boruff, B.J.; Shirley, W.L. Social vulnerability to environmental hazards. Soc. Sci. Quart. 2003, 84, 242–261.

58. Reid, C.E.; O'Neill, M.S.; Gronlund, C.J.; Brines, S.J.; Brown, D.G.; Diez-Roux, A.V.; Schwartz, J. Mapping community determinants of heat vulnerability. Environ. Health Perspect. 2009, 117, 1730–1736.

59. Cutter, S.L.; Mitchell, J.T.; Scott, M.S. Revealing the vulnerability of people and places: A case study of georgetown county, south Carolina. Ann. Assn. Amer. Geogr. 2000, 90, 713–737.

60. Minnesota Population Center. Integrated Public Use Microdata Series, International: Version 6.2 (Machine-Readable Database); University of Minnesota: Minneapolis, MN, USA, 2013.

61. Tran, K.; Azhar, G.; Nair, R.; Knowlton, K.; Jaiswal, A.; Sheffield, P.; Mavalankar, D.; Hess, J.J. A cross-sectional, randomized cluster sample survey of household vulnerability to extreme heat among slum dwellers in Ahmedabad, India. Int. J. Environ. Res. Public Health 2013, 10, 2515–2543.

62. Hahn, M.B.; Riederer, A.M.; Foster, S.O. The livelihood vulnerability index: A pragmatic approach to assessing risks from climate variability and change—A case study in mozambique. Global Environ. Change 2009, 19, 74–88.

63. Johnson, D.; Wilson, J.; Luber, G. Socioeconomic indicators of heat-related health risk supplemented with remotely sensed data. Int. J. Health Geogr. 2009, 8.

64. Uejio, C.K.; Wilhelmi, O.V.; Golden, J.S.; Mills, D.M.; Gulino, S.P.; Samenow, J.P. Intra-urban societal vulnerability to extreme heat: The role of heat exposure and the built environment, socioeconomics, and neighborhood stability. Health Place 2011, 17, 498–507.

65. Eisen, L.; Eisen, R.J. Using geographic information systems and decision support systems for the prediction, prevention, and control of vector-borne diseases. Annu. Rev. Entomol. 2011, 56, 41–61.

66. Peng, R.; Bobb, J.; Tebaldi, C.; McDaniel, L.; Bell, M.; Dominici, F. Toward a quantitative estimate of future heat wave mortality under global climate change. Environ. Health Perspect. 2011, 119, 701–706.

67. Knowlton, K.; Lynn, B.; Goldberg, R.A.; Rosenzweig, C.; Hogrefe, C.; Rosenthal, J.K.; Kinney, P.L. Projecting heat-related mortality impacts under a changing climate in the new york city region. Amer. J. Public Health 2007, 97, 2028–2034.

68. Sheffield, P.E.; Knowlton, K.; Carr, J.L.; Kinney, P.L. Modeling of regional climate change effects on ground-level ozone and childhood asthma. Amer. J. Prev. Med. 2011, 41, 251–257.

69. Lu, Y.; Zeger, S.L. On the equivalence of case-crossover and time series methods in environmental epidemiology. Biostatistics 2007, 8, 337–344.

70. Bell, M.L.; Dominici, F.; Samet, J.M. A meta-analysis of time-series studies of ozone and mortality with comparison to the national morbidity, mortality, and air pollution study. Epidemiology 2005, 16, 436–445.

71. Medina-Ramón, M.; Zanobetti, A.; Cavanagh, D.P.; Schwartz, J. Extreme temperatures and mortality: Assessing effect modification by personal characteristics and specific cause of death in a multi-city case-only analysis. Environ. Health Perspect. 2006, 114, 1331–1336.

72. Naumova, E.N.; Egorov, A.I.; Morris, R.D.; Griffiths, J.K. The elderly and water-borne cryptosporidium infection: Gastroenteritis hospitalizations before and during the 1993 milwaukee outbreak. Emerg. Infect. Dis. 2003, 9, 418–425.

73. Kovats, R.; Edwards, S.; Hajat, S.; Armstrong, B.; Ebi, K.; Menne, B. The effect of temperature on food poisoning: A time-series analysis of salmonellosis in ten european countries. Epidemiol. Infect. 2004, 132, 443–453.

74. Craun, G.F.; Calderon, R.L. Observational epidemiologic studies of endemic water-borne risks: Cohort, case-control, time-series, and ecologic studies. J. Water Health 2006, 4, 101–119.

75. Maclure, M. The case-crossover design: A method for studying transient effects on the risk of acute events. Amer. J. Epidemiol. 1991, 133, 144–153.

76. Biglan, A.; Ary, D.; Wagenaar, A.C. The value of interrupted time-series experiments for community intervention research. Prev. Sci. 2000, 1, 31–49.

77. Higgins, J.; Green, S. Cochrane Handbook for Systematic Reviews of Interventions; Version 5.1.0. The Cochrane Collaboration: Oxford, UK, 2011.

78. Huang, C.; Barnett, A.G.; Wang, X.; Vaneckova, P.; FitzGerald, G.; Tong, S. Projecting future heat-related mortality under climate change scenarios: A systematic review. Environ. Health Perspect. 2011, 119, 1681–1690.

79. Meehl, G.A.; Covey, C.; Delworth, T.L. The wcrp cmip3 multimodel dataset: A new era in climate change research. Bull. AmER. Meteorol. Soc. 2007, 88, 1383–1394.

80. Nakicenovic, N.; Swart, R. Intergovernmental Panel on Climate Change Special Report on Emissions Scenarios; Cambridge University Press: Cambridge, UK, 2000.

81. Sheridan, S.C.; Allen, M.J.; Lee, C.C.; Kalkstein, L.S. Future heat vulnerability in California, Part II: Projecting future heat-related mortality. Clim. Change 2012, 115, 311–326.

82. Hess, J.; McDowell, J.; Luber, G. Adaptive Management of Distinctly Climate Sensitive Health Threats: The Example of Extreme Heat. In Proceedings of the American Public Health Association Annual Conference, Washington, DC, USA, 2 November 2011.

83. Brownson, R.C.; Fielding, J.E.; Maylahn, C.M. Evidence-based public health: A fundamental concept for public health practice. Annu. Rev. Public Health 2009, 30, 175–201.

84. Bouzid, M.; Hooper, L.; Hunter, P.R. The effectiveness of public health interventions to reduce the health impact of climate change: A systematic review of systematic reviews. PLoS One 2013, 8.

85. Eriksson, C. Learning and knowledge-production for public health: A review of approaches to evidence-based public health. Scand. J. Public Health 2000, 28, 298–308.

86. Hess, J.J.; Tlumak, J.E.; Raab, K.K.; Eidson, M.; Luber, G. An evidence-based public health approach to climate change adaptation. Environ. Health Perspect.. under review.

87. Bassil, K.L.; Cole, D.C. Effectiveness of public health interventions in reducing morbidity and mortality during heat episodes: A structured review. Int. J. Environ. Res. Public Health 2010, 7, 991–1001.

88. Moher, D.; Liberati, A.; Tetzlaff, J.; Altman, D.G. Preferred reporting items for systematic reviews and meta-analyses: The prisma statement. Ann Intern Med. 2009, 151, 264–269.

89. Grosse, S.D.; Teutsch, S.M.; Haddix, A.C. Lessons from cost-effectiveness research for United States public health policy. Annu. Rev. Public Health 2007, 28, 365–391.
90. Institute of Medicine. Committee on Comparative Effectiveness Research Prioritization. Initial National Priorities for Comparative Effectiveness Research; National Academies Press: Washington, DC, USA, 2009.
91. Velentgas, P.; Dreyer, N.A.; Nourjah, P.; Smith, S.R.; Torchia, M.M. Developing a Protocol for Observational Comparative Effectiveness Research: A User's Guide; Government Printing Office: Washington, DC, USA, 2013.
92. Haddix, A.C.; Teutsch, S.M.; Corso, P.S. Prevention Effectiveness: A Guide to Decision Analysis and Economic Evaluation; Oxford University Press: Oxford, UK, 2003.
93. Eakin, H.C.; Tompkins, E.L.; Nelson, D.R.; Anderies, J.M. Hidden Costs and Disparate Uncertainties: Trade-offs in Approaches to climate Policy. In Adapting to Climate Change: Thresholds, Values, Governance; Adger, W.N., Lorenzoni, I., O'Brien, K.L., Eds.; Cambridge University Press: Cambridge, UK, 2009; pp. 212–226.
94. Successful Adaptation to Climate Change: Linking Science and Policy in a Rapidly Changing World; Moser, S.C., Boykoff, M.T., Eds.; Routledge: New York, NY, USA, 2013.
95. Climate Change, Ethics and Human Security; O'Brien, K., St. Clair, A.L., Kristoffersen, B., Eds.; Cambridge University Press: New York, NY, USA, 2010.
96. Natural Resources Canada. Climate Change Adaptation Planning: A Handbook for Small Canadian Communities; Natural Resources Canada: Ottawa, ON, Canada, 2011.
97. Margoluis, R.; Stem, C.; Salafsky, N.; Brown, M. Using conceptual models as a planning and evaluation tool in conservation. Eval. Program. Plan. 2009, 32, 138–147.
98. Health and Spatial Planning: Royal Town Planning Institute Policy Statement; Royal Town Planning Institute: London, UK, 2013.
99. PlaNYC. Available online: http://www.nyc.gov/html/planyc2030/html/home/home.shtml (accessed on 18 December 2013).
100. Sustainable Sydney 2030. Available online: http://www.cityofsydney.nsw.gov.au/vision/sustainable-sydney-2030 (accessed on 18 December 2013).
101. Ten Scenarios for "Grand Paris" Metropolis Now Up for Public Debate. Available online: http://www.bustler.net/index.php/article/ten_scenarios_for_grand_paris_metropolis_now_up_for_public_debate/ (accessed on 18 December 2013).
102. Greater London Authority. The London Plan. Available online: http://www.london.gov.uk/priorities/planning/london-plan (accessed on 18 December 2013).
103. Office of Planning Environment and Realty. Planning Horizons for Metropolitan Long-range Transportation Plans. Available online: http://www.fhwa.dot.gov/planning/planhorz.cfm (accessed on 18 December 2013).
104. American Water Works Association. Dawn of the Replacement Era: Reinvesting in Drinking Water Infrastructure. Available online: http://www.win-water.org/reports/infrastructure.pdf (accessed on 18 December 2013).
105. Climate Change: A Serious Threat to Public Health: Key Findings from Astho's 2009 Climate Change Needs Assessment; ASTHO: Arlington, VA, USA, 2009.

106. Association of State and Territorial Health Officials. Budget Cuts Continue to Affect the Health of Americans; Association of State and Territorial Health Officials: Arlington, VA, USA, 2013; p. 3.
107. Local Health Department Job Losses and Program Cuts: Findings from the 2013 Profile Study; NACCHO: Washington, DC, USA, 2013.

# CHAPTER 11

# The Los Angeles County Community Disaster Resilience Project: A Community-Level, Public Health Initiative to Build Community Disaster Resilience

DAVID EISENMAN, ANITA CHANDRA, STELLA FOGLEMAN,
AIZITA MAGANA, ASTRID HENDRICKS, KEN WELLS,
MALCOLM WILLIAMS, JENNIFER TANG, AND ALONZO PLOUGH

## 11.1 INTRODUCTION

As disasters increase in scale, frequency, length, and costs worldwide it is apparent that communities cannot rely on national governmental dollars and agencies to ensure effective and comprehensive disaster response and recovery. Also, disasters in urban centers with diverse communities and growing inequalities challenge governmental capabilities to handle the complex social, health, housing, and financial challenges of response and recovery without local, community involvement [1]. In preparation for the post-2015 Hyogo Framework for Action, countries across the globe

*The Los Angeles County Community Disaster Resilience Project—A Community-Level, Public Health Initiative to Build Community Disaster Resilience.* © *Eisenman D, Chandra A, Fogleman S, Magana A, Hendricks A, Wells K, Williams M, Tang J, and Plough A.* International Journal of Environmental Research and Public Health *11,8 (2014). doi:10.3390/ijerph110808475. Licensed under a Creative Commons Attribution 3.0 Unported License, http://creativecommons.org/licenses/by/3.0.*

emphasize that local governments and community organizations must be supported and encouraged to implement community resilience programs [2]. In the United States, local health departments and responder agencies have often turned to non-governmental agencies and local community and faith based organizations during disasters for their knowledge of needs, resources and social complexities in the neighborhoods they serve [3,4,5,6].

Community resilience, or the sustained ability of a community to withstand and recover from adversity, emphasizes that effective and efficient disaster risk reduction, response and recovery requires a whole of community approach, specifying that partnerships with nongovernmental partners, engagement of local communities and orientation to community self-sufficiency is the foundation of this approach [7]. The World Health Organization urges member states to use coordinated, multisectoral approaches to disaster risk reduction, response and recovery [8]. In the United States, community resilience has become integral to several national directives [9,10], including the Center for Disease Control (CDC) and Prevention's public health emergency preparedness (PHEP) cooperative agreements [11] and the Federal Emergency Management Agency's (FEMA) "Whole of Community Planning" imperative [12]. FEMA's approach stipulates as policy that collaboration, local empowerment and collective community response are central to preparedness, response and recovery [13]. "The CDC PHEP agreements that give funding to state and local health departments provide a set of public health preparedness capabilities to assist health departments with their strategic planning for improving preparedness and creating more resilient communities" [11]. Capability 1 specifically addresses community resilience in the area of community preparedness. It includes determining community health risks, identifying vulnerable or at-risk populations and those with access and functional needs, building community partnerships, engaging community organizations to foster public health social networks, and coordinating training to enhance engagement of lay persons in community resilience.

While these mandates support preparedness activities at the community level, there are few evidence based methods to building community resilience in the United States [14]. It is unclear what local public health should do or the role it should play to build community resilience [15]. Public health officials faced with operationalizing the CDC capabilities need both evidence-based strategies for implementing community disaster resilience and evaluation methods for measuring results.

This paper describes the theoretical rationale, intervention design and evaluation of a public health program for increasing community resilience (CR) in selected neighborhoods in a major, urban center. The Los Angeles County Community Disaster Resilience (LACCDR) Project is providing the opportunity to translate a theory of community resilience building into practice, strengthen disaster resilience in Los Angeles County communities, and to evaluate its outcomes. The goal of the LACCDR Project is to increase the readiness of communities and the people who live in them to prepare for, respond to, and recover from a natural disaster, other emergency, or public health event through a community-based approach. This paper is intended to provide public health and academic researchers with new tools to conduct their community resilience programs and evaluation research.

## 11.2 METHODS

### 11.2.1 INTERVENTION DEVELOPMENT

As previously reported, the LACCDR project was developed over a period of two years by the Los Angeles County Department of Public Health working closely with community, academic, government and business partners [16,17]. During 2010–2012, the project engaged a broad array of community stakeholders representing government agencies and community-based organizations to identify and develop strategies that would bolster resilience. The project engaged these stakeholders through community forums, working groups, and community surveys that led to the design of the LACCDR Project [17].

### 11.2.2 COLLABORATION WITH DIVERSE STAKEHOLDERS

The foundation of the LACCDR Project is collaborations between public health, several academic institutions (UCLA, RAND Corporation, Loma

Linda University, USA), governmental and non-governmental agencies (United States Geological Survey Science Application for Risk Reduction, Emergency Network of Los Angeles, USA), businesses (private media consultants) and the communities themselves. These stakeholders comprise the LACCDR Steering Committee that guides the implementation and evaluation of the project. The Project is funded by the CDC PHEP grant with supplemental funding provided by the National Institutes of Mental Health and the Robert Wood Johnson Foundation.

**TABLE 1:** Description of the Sixteen Communities and Coalitions in the Los Angeles County Community Disaster Resilience Project (LACCDR) Project.

| Community | Population | Race/Ethnicity | Median Household Income | Percent Renters | Participating Organizations |
|---|---|---|---|---|---|
| Acton & Agua Dulce | 10,938 | His/Lat:18.1% White: 75.9% Afr. Amer: 1.0% Asian: 2.0% | Acton: $87,896; Agua Dulce: $97,000 | 10.8% | Town Committees, Sheriff's Department, CERT, U.S. Forest Service |
| La Crescenta | 19,653 | His/Lat: 11.4%, White: 57.9%, Afr. Amer: 0.7% Asian: 27.2% | $83,048 | 35.6% | Fire Safe Committee, Chamber of Commerce, CERT, Fire and Sheriff's Departments, Assisting Seniors Through Enhanced Resources (ASTER) |
| Pomona | 149,058 | His/Lat: 70.5%, White: 12.5%, Afr. Amer: 6.8% Asian: 8.3% | $50,893 | 44.9% | Emergency Manager, Pomona College, Chamber of Commerce American Red Cross, City Youth and Family services, Police, Tri-city Mental Health |
| Pico Union | 44,664 | His/Lat: 66.4%, White: 9.1%, Afr. Amer: 6.2% Asian: 16.5% | $26,424 | 89.4% | Neighborhood Committee, Police, Fire, County School District, Elementary Schools, Neighborhood Watch, Prevencion y Rescate, Salvation Army, Health Center, Pueblo Nuevo, Kolping House, Latino Community Chamber of Commerce |

**TABLE 1:** *Cont.*

| Community | Population | Race/Ethnicity | Median Household Income | Percent Renters | Participating Organizations |
|---|---|---|---|---|---|
| Culver City | 38,883 | His/Lat: 23.2%, White: 48.0% Afr. Amer: 9.2% Asian: 14.5% | $75,596 | 45.7% | Culver City Coalition, Westside Children's Center, Sony Pictures Entertainment, City School District, Open Paths Counseling Center, Kids are 1st, Medical Center |
| Watts | 51,223 | His/Lat: 73.4%, White: 0.6% Afr. Amer: 24.8%, Asian: 0.2% | $25,161 | 60.0% | Watts Gang Task Force, Kaiser Permanente, City of LA, Concerned Citizens of South Central Los Angeles, Housing Authority, The Center of Grief and Loss |
| Huntington Park | 58,114 | His/Lat: 97.1%, White: 1.6%, Afr. Amer: 0.4% Asian: 0.6% | $35,107 | 73.0% | Community Development Corporation, Fire, Police, American Red Cross, Head Start, Salvation Army, Chamber of Commerce |
| Wilmington | 53,815 | His/Lat: 88.8%, White: 4.7% Afr. Amer: 2.7% Asian: 2.0% | $37,277 | 60.4% | Tzu Chi Clinic, Hubbard Christian Center, Chamber of Commerce, American Red Cross, G.A.P. (Gang Alternative Program), Veterans of Foreign Wars, The Wilmington Teen Center, Philips 66 Refinery, Women of Wilmington, Port Police |
| Quartz Hill | 10,912 | His/Lat: 24.6%, White: 62.3%, Afr. Amer: 6.9%, Asian: 2.6% | $56,070 | 30.4% | Town Committee, Fire, Sheriff's Department, County School District, Water Board, CERT |
| San Fernando | 23,645 | His/Lat: 92.5%, White: 5.3%, Afr. Amer: 0.6%, Asian 0.8% | $52,021 | 45.5% | City Parks and Recreation, State University Police Services, Valley Care Health System, Mission Community Hospital, Partners in Care Foundation, Police, Public Works, Providence Health Services |

**TABLE 1:** *Cont.*

| Commu-nity | Popula-tion | Race/Ethnicity | Median Household Income | Percent Renters | Participating Organizations |
|---|---|---|---|---|---|
| San Gabriel | 39,718 | His/Lat: 25.7%, White: 11.4%, Afr. Amer: 0.8%, Asian: 60.4% | $57,666 | 50.8% | La Casa de San Gabriel Community Center, Asian Youth Center, Fire, Hope Christian Fellowship, First Presbyterian, St. Anthony's, Church of our Savior |
| Holly-wood | 27,434 | His/Lat:: 32.1% White: 48.7% Afr. Amer: 7.6% Asian: 8.1% | $31,415 | 96.0% | Neighborhood Commit-tees, Hollywood United (H.U.N.K.), United Method-ist Church, American Red Cross |
| Palms | 57,964 | His/Lat: 29.7%, White: 36.8%, Afr. Amer:10.1%, Asian: 18.9% | $60,728 | 78.8% | Fire, Police Department, American Red Cross, Com-munity Police Advisory Board (CPAB) |
| Comp-ton | 96,455 | His/Lat: 65.0%, White: 0.8%, Afr. Amer:32.1%, Asian: 0.2% | $43,311 | 44.8% | PACRED churches, Sheriff's Department, Compton Uni-fied School District, Comp-ton Office of Emergency Management, YWCA. |
| Hawai-ian Gardens | 14,254 | His/Lat: 77.2%, White: 7.3%, Afr. Amer: 3.4%, Asian: 10.5% | $42,898 | 55.7% | Emmanuel Church, Celebra-tion Christian Center, Fire Department, City of Hawai-ian Gardens, School District, City Committeeman |
| Gardena | 58,829 | His/Lat: 37.7%, White: 9.3%, Afr. Amer: 23.9% Asian: 25.8% | $46,961 | 52.1% | South Bay Coalition for the Homeless, Police Depart-ment, CERT, Asian Commu-nity Center, Baptist Church |

## 11.2.3 INTERVENTION

The basic design of the LACCDR Project is a pretest–posttest with com-parison group design [18]. A list of candidate communities for the project was developed with the Los Angeles County Department of Public Health's Service Planning Area (SPA) leadership, and other community leaders. Through this process the Steering Committee identified communities that fit the following criteria: (1) Modest-size population, ideally under 50,000 persons; (2) Shared identity as a "community" with at least two of the following: local business community; school/school district; police and fire department services; community clinic/hospital/ health responsible entity; engaged community based organizations; (3) Sufficient community organizational infrastructure to lead local knowledge and capability development and implement LACCDR (i.e., via neighborhood planning group); and (4) Diversity in risk exposure and culture/ethnicity. This list was submitted to the Steering Committee who narrowed it down to 16 communities (2 per SPA) that were matched on demographic and hazard risk characteristics (Table 1). Finally, the team's statistician randomly assigned the two neighborhoods in each of the eight SPAs, for a total of 16 communities, to the intervention or control condition. This entire process lasted from September to December 2012.

The conceptual framework for the intervention comes from the work of Anita Chandra and colleagues [7], the Los Angeles County Department of Public Health, and community members. As illustrated in Figure 1, they identified eight essential levers for creating community disaster resilience: wellness, access, education, engagement, self-sufficiency, partnership, quality and efficiency. The LACCDR Project is structured on four of these levers: education, engagement, self-sufficiency, and partnership. Education ensures ongoing information about preparedness, risks and resources before, during, and after a disaster. Engagement involves including community members and promoting participatory decision making in planning, response and recovery activities. Self-sufficiency refers to enabling and supporting individuals and communities to assume responsibility for their preparedness. Organizational partnership involves increasing and enhancing the linkages and collaborations between government and non-governmental organizations (NGOs) and between NGOs in the community.

The Logic Model shown in Figure 2 guides the Project. The outcomes are a result of outputs including nurses trained in CR, a practical CR-building toolkit, and community-based activities led by the coalitions for building their communities' resilience. The outputs come from operationalizing the levers of resilience in the Chandra model (Engagement, Partnerships, Education, and Self-Sufficiency) through specified activities by the partners and community coalitions using inputs from CDC and funders, the Steering Committee themselves, LACDPH, and the community coalitions. The Outcomes listed in the Logic Model constitute metrics tracked and reported in the evaluation.

Communities randomly assigned to the community resilience group (CR) each have a public health nurse assigned to work with the existing local neighborhood organization to develop a CR coalition dedicated to improving their community's resilience following community engagement principles. The CR public health nurses are trained in community resilience topics. These topics are assembled into a Toolkit, which is a set of strategies and materials that the community coalitions can use to build resilience in their communities (Table 2). The toolkit developed based on findings from the LACCDR Development Phase includes components deemed a high priority by LA County stakeholders [17]. The modules include sections on how to map the hazards and methods in a neighborhood using web-based tools and other resources; identification of at-risk or vulnerable populations in the community; understanding how to identify and respond to social and psychological trauma; and understanding how to utilize CR trained field workers. The toolkit is designed to be interactive with questions and activities to generate community-specific discussions. For instance, in one module each coalition conducts a community hazard assessment exercise with the goal of prioritizing the hazards on which to focus.

A novel resource developed specifically for the LACCDR Project is mapping software called Sahana (Sahana Software Foundation) that allows communities to map their sources of risk and resilience [19]. Communities in the resilience arm have access to training in and use of mapping and charting software that allows them to visualize the relationships between local hazards, socio-demographic subsets of their community (with a focus on vulnerable populations), and assets and resources available to the community (community organizations, CERT trained individuals, evacuation routes). Through mapping, CR communities can map which hazards may affect their community to support planning and prioritization, assess the relationship between hazards and vulnerable populations, and incorporate the roles of community-based organizations, faith-based organizations, government agencies and other sectors that are local resources to be engaged in their resilience planning. Sahana is an open source disaster management software that features the capability to display and overlay maps, manage resources, register organizations, and conduct surveys. Hazard maps, census data, and resource datasets were developed and loaded into Sahana for each of the eight CR communities. Hazard maps were provided by the United States Geological Survey Science Application for Risk Reduction program. Resource data came from the Los Angeles County Resource Listing available on the Los Angeles County GIS Data Portal [20]. Sociodemographic data were downloaded from the American Community Survey 2010. Specific functionalities were built to allow community members to collect and input further data on hazards, organizations, and community events. The website was built with membership from the coalitions advising through a user's group. All nurse facilitators in the CR arm and members of the eight coalitions received training in using Sahana. Ongoing technical assistance is available from Sahana to the coalitions.

The coalitions meet monthly in their communities and the CR public health nurses lead their coalitions through training in the Toolkit, with technical assistance from Steering Committee members as requested. Based on this process, each CR coalition develops a written CR Workplan for improving community resilience in their neighborhood. The coalitions choose a yearlong scope of work in the CR Workplan. The CR Workplan is informed by their trainings in the Toolkit sections and knowledge of their local priorities and is intended to build on the assets and partnerships existing in the coalition and in the community. Using the CR model and toolkit sections is not required but is encouraged with direction and support from the public health nurses and Steering Committee members. The "Community Resilience Measure" created specifically for this project is a tool meant to facilitate the translation of the CR model into their CR Workplans. The Community Resilience Measure guides the coalitions in creating a plan that operationalizes CR principles from the Chandra model (see Table 3). Each coalition's CR Workplan is shared with the Steering Committee who provides a review based on the Community Resilience Measure and recommendations for enhancing the CR Workplan so that it addresses individual items in the Community Resilience Measure. Once the CR Workplan is reviewed and approved by LACDPH in conjunction with projected expenditures, the coalition receives $15,000 in funding to use in implementing their plan.

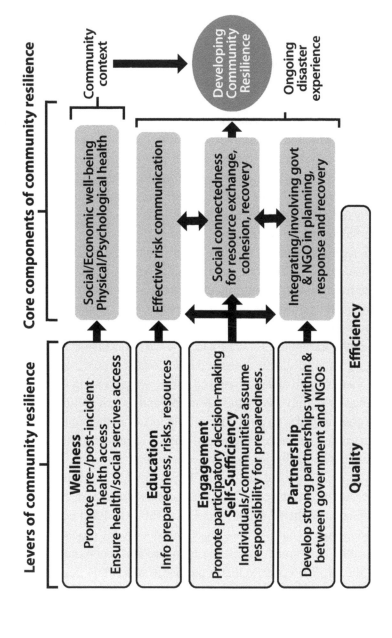

**FIGURE 1:** Conceptual Framework for LACCDR (reprinted with permission from Chandra et al. 2013) [7].

FIGURE 2: Logic Model for LACCDR.

| Inputs | CR Levers | Activities | Outputs | Outcomes (evaluation method) |
|---|---|---|---|---|
| LACDPH | Education | Committee builds toolkit, mapping website, Community Resilience Measure | Community Resilience Toolkit for dissemination | Increased organizational networks among Coalitions (PARTNER Tool) |
| Academic Partners | Engagement | Committee provides technical assistance | Community Resilience Measure for planning | Increased CR skills among coalitions (Table-top Exercise) |
| USGS | Self-sufficiency | Train nurses in CR | CR trained nurses | Increased participation in CR activities (PHRETS) |
| ENLA | Organizational Partnerships | Train communities in CR and Toolkit | Coalitions broaden membership from 11 sectors | Increased CR skills, such as PFA (PHRETS) |
| Media Consultant | | Coalitions write CR Workplan | CR activities implemented in community | Increased self-sufficiency (PHRETS) |
| Community coalitions | | Community media campaign | | Increased social networks available for disasters (PHRETS) |
| Funders | | | | |

**TABLE 2:** Description of Community Resilience Toolkit in LACCDR Project.

| Toolkit Section | Description (and Levers Addressed in Section) |
| --- | --- |
| Psychological First Aid | Psychological First Aid is designed to reduce disaster-induced stress by prompt provision of social support, linkage to resources, and promotion of effective coping strategies and coping self-efficacy. |
| | (Levers: Education, Self-Sufficiency) |
| Community Mapping | Community mapping is a process to identify resources and develop connections among people and their local organizations. There are several options for engaging in this process that vary in their use of technology, scope, and scalability. An aspect of the community mapping activity is helping communities consider access and functional needs populations. |
| | (Levers: Education, Self-Sufficiency, Engagement) |
| Community Engagement Principles for CR | A set of community engagement principles are applied to CR building initiatives and are applicable to responder agencies, community and faith-based organizations, community leaders and stakeholders, and community members. |
| | (Levers: Education, Self-Sufficiency, Engagement, Organizational Partnerships) |
| How to Identify and Develop Community Leaders | This section provides supports for communities to have effective leadership for CR. |
| | (Levers: Engagement, Organizational Partnerships) |
| Training Community Field Workers | Guidelines and resources for CR field workers, including nurses, school staff, and lay community health workers to support CR in communities. Includes a curricula on disaster preparedness. |
| | (Levers: Education, Self-Sufficiency, Engagement) |

The comparison group has a public health nurse or health educator assigned to develop a Preparedness coalition. In the comparison arm, the public health nurse or health educator takes a traditional educational approach by training the coalitions using a standardized, organized manual of public health practice for improving disaster preparedness [21]. (Disaster preparedness is conceptualized as focusing mainly on personal or household self-sufficiency through accumulation of supplies and emergency communication plans). Training topics include individual and family preparedness; considerations for special populations such as kids, animals, special needs, and seniors; communication tools; and linking with local nonprofits, faith-based organizations, and small businesses. The public health nurses or health educators who serve this function are not trained in community resilience and neither the public health workers nor the communities participating in the comparison arm have access to the CR toolkit or subject matter experts for technical assistance. Preparedness coalitions also create a written workplan and once approved, receive $15,000 to implement their activities.

**TABLE 3:** Community Resilience Measure.

| Thinking about Your Community's Plan Overall, Please Answer the Following Questions: |
| --- |
| **Priority Vulnerable Community Members (Levers addressed: Engagement, Organizational Partnerships)** |
| Who are your most vulnerable community members? |
| What mapping tool or other processes are you using to identify those vulnerable community members and where they are concentrated? |
| What are the limitations these vulnerable community members have in either mobility, communications or resources that make them particularly vulnerable in a disaster? |
| How are you including those vulnerable members in your planning process? (Planning "with" not "for" them.) |
| What are the assets, resources, and networks that vulnerable community members already have and how are you using them in your resilience plan? |
| **Understanding Your Community (Levers addressed: Engagement, Organizational Partnerships, Self-Sufficiency)** |
| What mapping tool or other process are you using to identify the hazards in your community? |
| What mapping tool or other process are you using to identify your community's resources? |
| How are you using the information you collected to get your neighbors and your community prepared, ready to respond, and able to recover from a disaster or emergency? |
| How are you encouraging neighbor to neighbor discussion or planning to support one another in a disaster or emergency? |
| **Important Sectors in your Community (Levers addressed: Engagement, Organizational Partnerships, Self-Sufficiency)** |
| How are you getting organizations and agencies from the CDC 11 sectors involved in the coalition? |
| List the types of organizations you've identified (those you already have, those you wish to still bring onboard)? |
| What roles do they play in the coalition (i.e., leading or supporting activities)? |
| How are you using the services and resources that these organizations and agencies bring in your community? |
| How are you coordinating the work of first responders and community members to avoid overlap and keep information flowing and lines of communication open? |

**TABLE 3:** *Cont.*

| Recovery (Levers addressed: Organizational Partnerships, Self-Sufficiency) |
| --- |
| How are you planning to help families, neighborhoods, and the community as a whole recover? |
| How will organizations and agencies in your community continue to help their current clients as well as the wider community, too? |
| How are organizations and agencies in your community involved in planning for the recovery process? |

## 11.3 EVALUATION

Evaluation of outcomes of the LACCDR Project aims to identify improvements in indicators of community resilience at multiple levels. The project uses a mixed methods evaluation strategy designed to focus on changes in community organization relationships, population practices and awareness, and evidence of change in the coalitions' skills and understanding of CR. These are being measured by an organizational network survey, a population-based survey, and table-top exercises with the coalitions, respectively. Specific descriptions follow below.

### 11.3.1 ORGANIZATIONAL NETWORK ANALYSIS

As discussed above, one of the theoretical levers for changing CR is increasing organizational partnerships by increasing and deepening the linkages among community NGOs. To assess improvements in this domain, the Project is measuring longitudinal changes in inter-organizational linkages among NGOs in the 16 communities. This evaluation seeks to determine if the coalitions in the community resilience arm improve their partnerships with community organizations, as measured by increases in their organizational linkages, compared to coalitions in the comparison group. The Project uses a social network analysis tool called PARTNER (Program to Analyze, Record, and Track Networks to Enhance Relationships) at the start of study and at least once more during the study period [22]. Data collected by the project will determine the quality of relationships among partners, how they change over time, and examine how they are leveraged to achieve resilience outcomes in eight communities in LA County. PARTNER is a software program that consists of a brief survey linked to an analysis tool that visually maps the collaborative network and analyzes the number, strength, and quality of connections among partners. This tool provides a means for measuring the process of exchange and interaction among participating organizations in a coalition over time and the activities in which each coalition is engaged.

Partnerships are measured primarily by connectivity. Connectivity is defined as the measured interactions between partner organizations in the coalition, such as the amount and quality of interactions, and ways in which these relationships change over time [22]. The following measures are used to assess connectivity: Types of relationship (e.g., is there information sharing, joint program development, or resource exchange?); trust among partners (measured as an index of reliability, mission agreement, and ability to have open discussion around issues); and value of the partner organizations to the coalition's mission (measured as power/influence, commitment, and resources provided).

Two levels of community resilience outcomes will also be measured, intermediate capacities of the coalitions and their final outcome capabilities. Intermediate capacities are necessary for the coalition to work together to achieve its goals of improving community resilience. Coalition capacities include having completed an assessment of vulnerabilities and assets, having a plan for delivering psychological first aid in an emergency, and recruiting volunteers for coalition activities. In addition, coalitions should impact community resilience capability outcomes. These outcomes describe the capacity of the system to provide both routine and emergency services. Capabilities include coalition members participating in exercises and drills, the community exercising its plan in an actual emergency, and the community has closer ties to the Los Angeles County Department of Public Health.

## 11.3.2 PHRETS HOUSEHOLD SURVEY

An outcome indicator of the Education, Self-Sufficiency and Engagement levers is longitudinal changes in neighborhood residents' community resilience activities and attitudes. A survey measures the resilience-related awareness, attitudes and practices of residents in the sixteen Los Angeles County communities. The survey is a repeated cross-sectional design conducted in English, Spanish and Korean languages.

Because the research design involves distinct treatment and control communities it is essential to select a sample of each of the sixteen communities, and for each sample element to be geographically linked to one and only one of those sixteen communities. For that reason, address-based sampling is being used. A sample size of $N = 4400$ participants at each time point ($n = 2200$ per arm) provides enough power to detect 4.22% change in two proportions without adjusting for clustering and 8.74% change in two proportions adjusting for clustering ($ICC = 0.01$). Survey domains include: household preparedness for disaster; participation in community resilience building activities; self-efficacy for helping in a disaster; perceived collective efficacy of the community in a disaster; perceived benefits of individual preparedness and perceived benefits of disaster planning with neighbors; locus of responsibility; trust in public health in a disaster; social networks available in a disaster; civic engagement; social cohesion; self-reliance in a disaster; perceived health and activity limitations; and demographics. The survey domains were selected as outcome indicators for the theoretical levers. For example, outcomes in the self-sufficiency domain can be measured by changes in household preparedness, self-efficacy for helping in a disaster, and perceived collective efficacy of the community in a disaster. Outcomes in the engagement domain can be measured by changes in participation in community resilience building activities. Outcomes in the education domain can be measured by changes in perceived benefits of individual preparedness and perceived benefits of disaster planning with neighbors.

## 11.3.3 TABLE-TOP EXERCISE

While the organizational network and community resident surveys are key to understanding practice, attitudes, and potential changes in resilience approaches, it is difficult to assess how a community or coalition may act in an actual event. In order to simulate an event condition and assess the extent to which coalitions were strengthening the four CR levers (engagement, self-sufficiency, partnership, and education), the Project developed and will conduct a series of CR tabletop exercises. The CR tabletop is built on a traditional tabletop design, used for events such as pandemic influenza [23], but is unique in its testing of how coalitions will work together to leverage community assets, address the needs of vulnerable populations, integrate government and NGO roles and plans, and ensure community ability to recover over the long-term. In order to test resilience principles, the table-top employs a scenario that is seemingly modest at start (a heat wave) but then escalates over time with other changes in community conditions (crime increases, drought worsens, brown-outs occur, and community members die). As designed, this allows the coalition to consider the extent and quality of their partnerships and assets for an expansive and lengthy event rather than what is traditionally tested in tabletops, that is, a massive, acute scenario.

The tabletop is designed to be relatively brief at two hours. The presentation of the scenario with prompts and two unfolding situations lasts 1.5 h. The debriefing and discussion that follow take 30 min. Participants are asked to rate how they responded during the scenario along the four levers. For example, participants are asked to rate their response on a scale of 1–5 for partnership, with 1 indicating that they have very little awareness of the sectors to bring into planning and response, and 5 noting all sectors are engaged and fully integrated into the response and recovery plan and that government and NGO is working collaboratively. A similar scale is used for the other three levers. In addition to coalition member ratings, the research team provides their own independent ratings based on observations. After the coalition concludes the tabletop, the study team provides a brief summary of their discussion. For the purpose of evaluation, the same tabletop exercise is conducted across all sixteen coalitions. However, the CR coalitions receive a more expansive summary with recommendations for action steps to improve their resilience responses (e.g., ideas and strategies for improving partnerships, ideas for considering the assets they need for recovery, considerations for the psychological and social impacts of long-term recovery), while the comparison coalitions only receive a brief summary of their discussion with no recommendations or insights from the study team. The study team then works with each CR coalition to consider the gaps or areas for improvement identified in the tabletop, with the goal of addressing those gaps over the next study year. The tabletop will be administered at two time point to assess change over time.

## 11.3.4 PROCESS EVALUATION

The project is also conducting a process evaluation using a mix of qualitative and quantitative methods (ethnography, participant reflections, document review) and measures of reach into the community. The process evaluation is examining the factors that promote development and impact of community coalitions as agents of change towards resilience. The process evaluation aims to describe how coalitions adopt resilience based principles and develop and implement community disaster resilience plans, including facilitators and barriers to these processes. During the coalition meetings, the evaluation team collects data that documents what the coalitions are doing and to what extent they are moving forward with respect to the four levers of resilience in the Chandra model. Observers attend meetings and take notes to capture how the 16 coalitions have: interpreted, adopted and adapted the information they have received; addressed the need for engagement of organizations and vulnerable populations from outside of their coalitions; discussed how to establish robust partnerships with other organizations/sectors within their communities; contemplated how to leverage their existing resources in efficient ways; and, considered how to plan at the community level rather than at the individual level. In addition, participants are asked to provide feedback through "Reflection Sheets" on successful and challenging elements of the meeting and toolkit training, raise any questions that they have about the Project, and suggest additional issues that they want addressed in the future.

## 11.4 DISCUSSION AND CONCLUSIONS

The LACCDR Project will shed light on several issues in the effort to build community resilience. We will have gained experience using community engagement and encouraging governmental and non-governmental partnerships as an approach to increasing resilience. We will learn more about training public health nurses in community resilience approaches, explaining resilience to community members, and what technical assistance is needed to be successful. We may learn how existing expertise and resources in our department of public health, collaborating organizations, and community coalitions can be leveraged to develop effective strategies that build communities that are more resilient to disasters. Finally, we will have data on the outcomes of a resilience building initiative and how it differs from the outcomes of a traditional method that focuses solely on household preparedness and does not actively promote community engagement and broad partnerships. The tools and measures we developed will be useful to similar efforts across the nation even as every community differs in leadership structure, assets, and other situational variations that require a different application of the tools we have developed and tested.

The LACCDR Project has limitations and challenges. A major limitation is constraints on measuring community resilience, a problem well known to practitioners in the field [24]. Project findings must be treated with caution as a result. A major challenge is that public health lacks tested tools for building community resilience. The process evaluation will provide important information for improving the toolkit so we expect it to be revised along the way. There will be variability among the communities in how they respond to the program because of differences in assets, resources, partnerships and leadership. Since much of community resilience is reflected in these differences, some degree of improved community resilience outcomes should be expected in both arms of the study.

The LACCDR Project is unusual in the United States because it implements and evaluates a public health led program for increasing community disaster resilience. Community resilience building activities can improve overall social cohesion and important aspects of community well-being so the implications of this study extend beyond the disaster preparedness area. It is therefore useful as a model for operationalizing policy directives for improving community disaster resilience and improving general community well-being. The study suggests specific programmatic activities and partnership approaches that can be implemented and evaluated in communities across the country.

The LACCDR Project is one of several such efforts to improve community resilience in the United States, most of which are led by first responder agencies such as the Federal Emergency Management Agency (FEMA) and the American Red Cross. America's PrepareAthon, led by FEMA, is a community-based program that aims to increase community engagement in community resilience planning and improve participants' knowledge of hazards and how to stay safe and mitigate damage [25]. American Red Cross is working on community resilience in many of its programs, some of which include community engagement as a component) [26].

However, neither of these efforts employ a systematic assessment of both individual and organizational change. Using findings from the LACCDR Project coupled with these other ongoing activities, state and local public health officers will soon have a more complete roadmap of how to translate high-level policy directives for community disaster resilience building into the implementation and evaluation of resilience building activities in communities.

## REFERENCES

1.  Baker, J.L. Climate Change, Disaster Risk, and the Urban Poor: Cities Building Resilience for a Changing World; World Bank Publications: Washington, DC, USA, 2012.
2.  Synthesis Report: Consultations on a Post-2015 Framework on Disaster Risk Reduction (HFA2); United Nations International Strategy for Disaster Reduction: Geneva, Switzerland, 2013.
3.  Kapucu, N. Collaborative emergency management: better community organising, better public preparedness and response. Disasters 2008, 32, 239–262.
4.  Joshi, P. Faith Based and Community Organizations' Participation in Emergency Preparedness and Response Activities; Institute for Homeland Security and Solutions Research: Triangle Park, NC, USA, 2010.
5.  Chandra, A.; Acosta, B. The Role of Nongovernmental Organizations in Long-Term Human Recovery after Disaster: Reflections from Louisiana Four Years after Hurricane Katrina; RAND Corporation: Santa Monica, CA, USA, 2009.
6.  National Research Council (U.S.). Committee on Private-Public Sector Collaboration to Enhance Community Disaster Resilience; National Academies Press (U.S.). In Building Community Disaster Resilience through Private-Public Collaboration; National Academies Press: Washington, DC, USA, 2011.

7. Chandra, A.; Williams, M.; Plough, A.; Stayton, A.; Wells, K.B.; Horta, M.; Tang, J. Getting actionable about community resilience: The Los Angeles County Community Disaster Resilience project. Amer. J. Public Health 2013, 103, 1181–1189.

8. Strengthening National Health Emergency and Disaster Managment Capacities and Resilience of Health Systems; World Health Organization: Geneva, Switzerland, 2011.

9. National Security Strategy; The White House: Washington, DC, USA, 2010.

10. National Health Security Strategy of the United States of America; United States Department of Health and Human Services: Washington, DC, USA, 2009.

11. Public Health Preparedness Capabilities: National Standards for State and Local Planning; U.S. Centers for Disease Control and Prevention: Atlanta, GA, USA, 2011.

12. National Disaster Recovery Framework: Draft; Federal Emergency Management Agency: Washington, DC, USA, 2010.

13. A Whole Community Approach to Emergency Management: Principles, Themes, and Pathways for Action; Federal Emergency Management Agency: Washington, DC, USA, 2011.

14. Schoch-Spana, M.; Courtney, B.; Franco, C.; Norwood, A.; Nuzzo, J.B. Community resilience roundtable on the implementation of Homeland Security Presidential Directive 21 (HSPD-21). Biosecur. Bioterror. 2008, 6, 269–278.

15. Schoch-Spana, M.; Sell, T.K.; Morhard, R. Local health department capacity for community engagement and its implications for disaster resilience. Biosecur. Bioterror. 2013, 11, 118–129.

16. Plough, A.; Fielding, J.E.; Chandra, A.; Williams, M.; Eisenman, D.; Wells, K.B.; Law, G.Y.; Fogleman, S.; Magaña, A. Building community disaster resilience: Perspectives from a large urban county department of public health. Amer. J. Public Health 2013, 103, 1190–1197.

17. Wells, K.B.; Tang, J.; Lizaola, E.; Jones, F.; Brown, A.; Stayton, A.; Williams, M.; Chandra, A.; Eisenman, D.; Fogleman, S.; Plough, A. Applying community engagement to disaster planning: developing the vision and design for the Los Angeles County Community Disaster Resilience initiative. Amer. J. Public Health 2013, 103, 1172–1180.

18. Campbell, D.T.; Stanley, J.C. Experimental and Quasi-Experimental Designs for Research; Rand McNalley: Chicago, IL, USA, 1966.

19. Sahana. Community Resilience Mapping Tool Demo. Available online: http://demo.lacrmt.sahanafoundation.org (accessed on 15 June 2014).

20. Los Angeles County GIS Data Portal. Available online: http://egis3.lacounty.gov/dataportal (accessed on 15 June 2014).

21. County of Los Angeles Emergency Survival Guide; Los Angeles County Chief Executive Office: Los Angeles, CA, USA, 2013.

22. Varda, D.M.; Usanov, A.; Chandra, A.; Stern, S. PARTNER (Program to Analyze, Record and Track Networks to Enhance Relationships); RAND Corporation: Santa Monica, CA, USA, 2008.

23. Dausey, D.J.; Aledort, J.E.; Lurie, N. Tabletop Exercise for Pandemic Influenza Preparedness in Local Public Health Agencies; RAND Corporation: Santa Monica, CA, USA, 2006; p. 67.

24. Disaster Resilience: A National Imperative; The National Academies Press: Washington, DC, USA, 2012.

25. The National Preparedness Community. America's PreparAthon! Available online: http://www.community.fema.gov/connect.ti/AmericasPrepareathon (accessed on 15 June 2014).

26. Robert Wood Johnson Foundation. Preparedness Summit: American Red Cross Community Resilience Pilot Program. Available online: http://www.rwjf.org/en/blogs/new-public-health/2012/02/preparedness-summit-american-red-cross-community-resilience-pilot-program.html (accessed on 15 June 2014).

# Author Notes

## CHAPTER 1

### Conflict of Interest

The authors declare no conflict of interest.

## CHAPTER 2

### Conflict of Interest

The authors declare no conflict of interest.

## CHAPTER 3

### Acknowledgments

We thank Karen Kerner for her technical and editorial review of our work and this paper. We also thank the peer reviewers for their insightful questions, which helped us improve the paper.

### Conflict of Interest

The authors declare no conflict of interest.

## CHAPTER 4

### Acknowledgements

We are grateful to the funding from the Canada Research Chairs program of the Social Sciences and Humanities Research Council (SSHRC) that has made this research possible.

## CHAPTER 5

### Acknowledgments
The author is heartily grateful for the team work and inspiration provided by all research team members in the joint research project titled "Social-ecological system resilience and adaptive co-management: an experimental pioneer study in Taiwan". This article presents part of the results of a subproject titled "Social-ecological system resilience and adaptive co-management: origins, interplay and changes of resilience and governance institutions", and financed by research grants from National Dong Hwa University (101T927-3; 102T929-5; 103T929-5) and the Ministry of Science and Technology, Taiwan (NSC 102-2621-M-259-001; MOST 103-2621-M-259-004). I thank these two institutions for their support. Special thanks are given to all interviewees, including community residents, government officials, non-governmental organization practitioners, and scholars, for their generous supports and feedback. I am also heartily grateful for the two anonymous reviewers for their insightful comments and suggestions.

### Conflict of Interest
The author declares no conflict of interest.

## CHAPTER 6

### Conflict of Interest
The authors declare no conflict of interest.

## CHAPTER 9

### Acknowledgments
This work was partially funded by Oak Ridge National Laboratory through subcontract No. 4000070670.

### Conflict of Interest
The author declares no conflict of interest.

## CHAPTER 10

### Acknowledgments
The authors gratefully acknowledge internal reviewers at the Centers for Disease Control and Prevention for their assistance in strengthening the manuscript.

### Author Contributions
Gino D. Marinucci, Jeremy J. Hess, and George Luber conceived of the manuscript, and all authors participated in drafting and revising the manuscript.

### Conflict of Interest
The authors declare no conflict of interest.

## CHAPTER 11

### Acknowledgments
This work was supported by grants from the Centers for Disease Control and Prevention (grant 2U90TP917012-11), the National Institutes of Health (grant P30MH082760; funded by the National Institute of Mental Health), and the Robert Wood Johnson Foundation. We thank all project staff and Steering Committee members for their contributions and commitment to the project.

### Author Contributions
David Eisenman conceptualized and led the writing of this paper. All of the other authors provided input and contributed to drafts and editing.

### Conflict of Interest
The authors declare no conflict of interest.

# Index